T0201113

Wetland Ecosystems

William J. Mitsch
The Ohio State University

James G. Gosselink
Louisiana State University

Christopher J. Anderson
Auburn University

Li Zhang
The Ohio State University

WILEY

John Wiley & Sons, Inc.

This book is printed on acid-free paper. ∞

Copyright © 2009 by John Wiley & Sons, Inc. All rights reserved.

Published by John Wiley & Sons, Inc., Hoboken, New Jersey
Published simultaneously in Canada

No part of this publication may be reproduced, stored in a retrieval system, or transmitted in any form or
by any means, electronic, mechanical, photocopying, recording, scanning, or otherwise, except as
permitted under Section 107 or 108 of the 1976 United States Copyright Act, without either the prior
written permission of the Publisher, or authorization through payment of the appropriate per-copy fee to
the Copyright Clearance Center, 222 Rosewood Drive, Danvers, MA 01923, (978) 750-8400, fax (978)
646-8600, or on the web at www.copyright.com. Requests to the Publisher for permission should be
addressed to the Permissions Department, John Wiley & Sons, Inc., 111 River Street, Hoboken, NJ
07030, (201) 748-6011, fax (201) 748-6008, or online at www.wiley.com/go/permissions.

Limit of Liability/Disclaimer of Warranty: While the publisher and the author have used their best efforts
in preparing this book, they make no representations or warranties with respect to the accuracy or
completeness of the contents of this book and specifically disclaim any implied warranties of
merchantability or fitness for a particular purpose. No warranty may be created or extended by sales
representatives or written sales materials. The advice and strategies contained herein may not be suitable
for your situation. You should consult with a professional where appropriate. Neither the publisher nor
the author shall be liable for any loss of profit or any other commercial damages, including but not limited
to special, incidental, consequential, or other damages.

For general information about our other products and services, please contact our Customer Care
Department within the United States at (800) 762-2974, outside the United States at (317) 572-3993 or
fax (317) 572-4002.

Wiley also publishes its books in a variety of electronic formats. Some content that appears in print may
not be available in electronic books. For more information about Wiley products, visit our web site at
www.wiley.com.

Library of Congress Cataloging-in-Publication Data:

Wetland ecosystems / William J. Mitsch ... [et al.].
     p. cm.
  Includes bibliographical references and index.
  ISBN 978-0-470-28630-2 (cloth)
  1. Wetland ecology—United States. 2. Wetland management—United States.
  3. Wetlands—United States. I. Mitsch, William J.
  QH541.5.M3W4645 2009
  577.68—dc22

                                                                    2009001797

Printed in the United States of America.
SKY10046349_050423

Contents

Preface

We are proud to introduce *Wetland Ecosystems*, which is both a stand-alone primer to a systems view of wetlands and a supplement to *Wetlands, 4th Edition* (Mitsch and Gosselink, 2007).

This new book emphasizes the structure and function of wetland ecosystems found in North America, and how we study them as systems. The book also uses ecosystem papers of wetland types from throughout the world to broaden its international coverage. The *raison d'etre* of this new book requires an explanation. The authors of *Wetlands 4th Edition* (here after referred to as *Wetlands 4*) removed the seven "ecosystem" chapters from previous editions to make the book more manageable in size for upper level/graduate courses in universities; we reduced the book length by 35 percent and streamlined it into two major sections—wetland science and wetland management. By its 3rd edition, the Mitsch and Gosselink wetlands textbook (*Wetlands 3*) had become encyclopedic; there was just too much material in it to cover effectively in one quarter/semester in a university course. In order to include a more complete description of wetland science (hydrology, biogeochemistry, biological adaptations, etc.) and wetland management we, with some trepidation, had to sacrifice the coverage of wetlands as ecosystems.

This new book, *Wetland Ecosystems*, is a supplement to that text, emphasizing the way we view wetlands that has served us so well—as ecosystems.

Wetland Ecosystems has a simple outline of 5 chapters. Chapter 1 gives a brief introduction to the definitions of both ecosystems and wetlands and a brief introduction with words, pictures, and energy flow diagrams to the main wetland ecosystems. The seven "lost" ecosystem chapters from *Wetlands 3* have been thoroughly updated and streamlined into three chapters—*Coastal Wetlands* (Chapter 2), *Freshwater Swamps and Marshes* (Chapter 3), and *Peatlands* (Chapter 4). These three chapters keep the same format of the ecosystem chapters of *Wetlands 1* (Mitsch and Gosselink, 1986), *Wetlands 2* (Mitsch and Gosselink, 1993) and *Wetlands 3* (Mitsch and Gosselink,

2000), first describing wetland ecosystem structure and then wetland ecosystem function. At least 80 new citations from hundreds that were reviewed are included in these chapters, providing new information on many topics—salt marsh die-off; invasive plants in tidal wetlands; tidal wetland resilience to climate change; ecology of tidal freshwater swamps; amphibian requirements in freshwater marshes; peatland methane emissions; and effects of climate change on peatlands—not covered extensively or at all in *Wetlands 3*.

Chapter 5, *Ecosystem Approaches to Wetland Science*, is new and covers how we study wetland ecosystems as systems. More than 100 new publications were cited in this chapter alone, with an emphasis on wetland ecosystem "models" of different time and space scales—1 m^2 mesocosms, multi-hectare experimental wetlands, and mathematical descriptions of wetlands ranging from 1 m^2 of marshland to a 10,000-km^2 Florida Everglades spatial model. This chapter describes mesocosm experiments of coastal wetlands in Australia, Denmark, and Louisiana, freshwater marshes in Ohio and Florida, and peatlands in Minnesota. Full-scale wetland experiments are then presented for such classic wetland research sites as the University of Florida cypress domes in north central Florida; the Delta Marsh in Manitoba; the Des Plaines River Wetland Demonstration Project in Illinois; the Olentangy River Wetland Research Park in Ohio; and the Bois-des-Bel Peatlands in Quebec. Overall, the book has almost 200 new references, all providing new information on wetland ecosystems, and most since 2000, which were not available or cited in either *Wetlands 3* or *Wetlands 4*.

We suggest that this book could be a supplement in upper-level wetland ecology university courses that use *Wetlands 4* as the main textbook. It can also be a stand-alone book that might be much more suitable for lower-level wetland ecology courses where the students do not have as much background in biology, chemistry, and physics. And of course, it remains a reference book for wetland ecosystems of North America and how we study them as ecosystems.

We would like to thank the following people for their help in putting together this book. Ruthmarie Mitsch did her usual heroic text editing and provided much assistance with citations. Anne Mischo continued her magic with illustrations, especially in developing a number of new and innovative illustrations in Chapter 5. Photos and other assistance were kindly supplied by Andy Baldwin (University of Maryland), Line Rochfort, Stéphanie Boudreau, and Claire Boismenu (Peatland Ecology Research Group, Université Laval, Québec), Ülo Mander and Valdo Kuusenets (Tartu University and University of Life Sciences, Estonia), Glenn Gutenspergen (U.S. Geological Survey, Patuxent Wildlife Research Center), Clay Rubec (Environment Canada), Erik Kristensen (University of Southern Denmark), Karen McKee (U.S. Geological Survey, National Wetlands Research Center, Lafayette, LA), Scott Bridgham (University of Oregon), Mike Waddington (McMaster University, Ontario), and Tom Oulette.

We also appreciate the support and encouragement offered by John Wiley & Sons, Inc., particularly Dan Magers, Kerstin Nasdeo, and Jim Harper. As always, we are pleased to work with the Wiley brand.

William J. Mitsch, Ph.D.
Columbus, OH

James G. Gosselink, Ph.D.
Rock Island, TN

Christopher J. Anderson, Ph.D.
Auburn, AL

Li Zhang, Ph.D.
Columbus, OH

March 2009

Introduction

This book is about wetland ecosystems. The operative root word is "systems." Ecosystem, as a term, was coined in the 1930s by Sir Arthur Tansley (1871–1955) and Roy Clapham (1904–1990), both British botanists. Tansley (1935) defined ecosystems as "the whole system including not only the organism-complex, but also the whole complex of physical factors forming what we call the environment of the biome—the habitat factors in the widest sense." As described by ecological historian Robert McIntosh (1985), Tansley's concept of ecosystem was picked up by Lindeman (1942) in his famous and, at the time, controversial study of trophic dynamics in Cedar Bog Lake (a peatland) in Minnesota. (This study is discussed further in Chapter 4, "Peatlands.") Thus, ecosystem ecology was born in wetlands, which may have provided the system for its initial incubation.

A more modern definition of "ecosystem" that we prefer, similar to the Tansley description, is *a complex of ecological communities and their environment, forming a functioning whole in nature* (reworded from Patten and Jørgensen, 1995). Most important to note is that an ecosystem includes the biological communities and the abiotic environment in which they are found. In some cultures, ecosystems are not recognized. For example, in Russia, the term used to define a similar concept is *biogeocoenosis*.

Ecosystems have been considered by many to be the most fundamental unit of study in ecology (McNaughton and Wolf, 1973; Patten and Jørgensen, 1995) and there is still the belief that the entire ecosystem needs to be studied to determine the importance of any one species or community within that ecosystem. We would like to think that is the case today, but, in fact, ecosystems are rarely studied as a whole. Rather, ecologists or teams of ecologists focus on specific parts of the ecosystem—vegetation, animal species, or groups of organisms (communities)—but less commonly include the abiotic environment or describe the interactions between the biotic and abiotic parts of the ecosystem. One exception to this is the investigation of biogeochemical cycles in wetlands, probably because the effects of the abiotic environment—hydrology, soils, and so on—are so obvious and important.

Wetland Ecosystems

This book is specifically about wetland ecosystems. Defining wetlands precisely is even more complex than defining ecosystems, although, properly, most wetland definitions include both the biotic (usually vegetation) and abiotic (soils and hydrology) components. Our companion textbook (Mitsch and Gosselink, 2007) gives seven definitions of wetlands, five of which come from the United States. We repeat a diagram from that book as our working definition of wetlands (see Figure 1-1). Wetlands are shown to be a three-component ecosystem. The hydrology of the landscape influences and changes the physiochemical environment, which in turn, along with hydrology, determines the biotic communities that are found in the wetland. Overall, the forcing functions of the wetlands, or any landscape ecosystem for that matter,

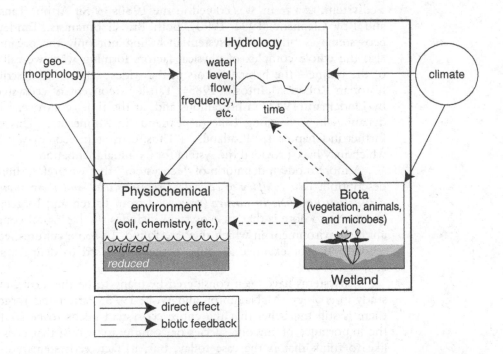

Figure 1-1 Conceptual model of a wetland ecosystem, showing the three-component basis of a wetland often used in wetland definitions, and the principal cause of wetlands—climate and landscape geomorphology. (*From Mitsch and Gosselink, 2007, 2000; NRC, 1995*)

include climate, which includes solar energy, temperature patterns, and precipitation. Climate couples with the geomorphology of the landscape to influence where and when water is present long enough to cause wetlands to exist.

Several wetland classification systems are used by scientists around the world to categorize wetland ecosystems. Some are simple, such as the Circular 39 system first used in the United States in the 1950s (Shaw and Fredine, 1956) with 20 wetland types and the system used internationally by the Ramsar Convention (Mitsch and Gosselink, 2007) with 27 wetland types. The detailed hierarchical system developed in the 1970s in the U.S. (Cowardin et al., 1979) as part of the U.S. National Wetlands Inventory and the more recent (1990s) hydrogeomorphic (HGM) classification system developed by Brinson (1993) in an attempt to include hydrodynamics in the classification of wetlands, are formal and all encompassing but are much too complex to use as way to organize this book. These and other formal wetland classifications are described in more detail in Chapter 8 of *Wetlands, 4th edition* (Mitsch and Gosselink, 2007).

In this book we describe wetlands divided into three major groups (see Table 1-1):

- Coastal wetlands—salt marshes, tidal freshwater marshes, mangrove swamps
- Freshwater swamps and marshes
- Peatlands

These classes of wetlands are generally recognizable ecosystems for which extensive research literature is available. Regulatory agencies also deal with these wetland system types, and management strategies and regulations have been developed for them. Collectively, these categories encompass most if not all of the 2.5 million km^2 of wetlands of North America and 6 to 8 million km^2 of wetlands in the world as a whole (Mitsch and Gosselink, 2007; Table 1-1).

Table 1-1 Wetland Ecosystem Types Described in This Book, with Their Estimated Area (x 10⁶ ha) in the United States, Canada, and the World

Type of Wetland	Wetland area, $\times 10^6$ ha			Book Chapter
	USA	Canada	World	
Coastal Wetlands				
Tidal Salt Marshes	1.9	1[a]	10[a]	2
Tidal Freshwater Marshes	0.8	—	2[a]	2
Mangrove Wetlands	0.5	—	24	2
Inland Wetlands				
Freshwater Marshes	27	16	95	3
Freshwater Swamps and Riparian Forests	25	—	109	3
Peatlands	55	110	350	4
TOTAL	110	127	580	

[a]Estimated.

How to Read These Ecosystem Diagrams

The energy flow diagrams that follow in this chapter and are found scattered through the remaining chapters in this book are full of information. They are drawn in the energy language, or "energese," developed by the late H. T. Odum (1924–2002). See http://en.wikipedia.org/wiki/Howard_T._Odum. A more appropriate term for the language might be "ecosystem language." The symbols were developed to describe energy flow in ecosystems but are now used to describe the dynamics of any system, from wetland to watershed to the biosphere. The symbolic energy language has been called the "shorthand for ecology," and scientists who are familiar with the language actually "talk" energese as they draw and redraw system diagrams until they agree that they have described how the systems work. Key references for the symbols are Odum and Odum (2000). The symbols are defined in Appendix A.

Coastal Wetlands

Several types of wetlands in the coastal areas are influenced by alternate floods and ebbs of oceanic tides. Near coastlines, the salinity of the water approaches that of the ocean, whereas further inland, the tidal effect can remain significant even when the salinity approaches that of freshwater. Coastal wetlands include tidal salt marshes, tidal freshwater wetlands, and mangroves swamps. We estimate that there are 0.36 million km^2 of coastal wetlands in the world (refer to Table 1-1). The area is not well known because of different definitions of "coastal" and because non-vegetated flats are a common type of coastal wetland in many regions and are counted in some inventories but not counted in others. The total area of wetlands considered as coastal or estuarine in the United States, including Alaska, is approximately 3.2 million ha, with about 1.9 million ha as salt marshes and 0.5 million ha as mangrove swamps. Alaskan estuarine wetlands alone are estimated to cover 0.9 million ha with only about 17% of these wetlands vegetated and thus presumably salt marsh (Hall et al., 1994). The vast majority of estuarine wetlands in Alaska were classified by Hall et. al. (1994) as "nonvegetated" or mudflats.

Salt marshes (see Figure 1-2) are found throughout the world along protected coastlines in the middle and high latitudes. As Figure 1-2 suggests, salt marshes are primarily detrital-based, with abundant fauna dependent directly (e.g., crabs) or indirectly (e.g., birds, estuarine fish) on this detrital production. A modest amount of the marsh grass productivity is consumed by grazing. Plants and animals in these systems have adapted to the stresses of salinity, periodic inundation, and extremes in temperature. One of the areas where salt marshes are most prevalent in the world is along the eastern coast of North America from New Brunswick to Florida and on to Louisiana and Texas along the Gulf of Mexico. Salt marshes are also found in narrow belts on the west coast of the United States and along much of the coastline of Alaska

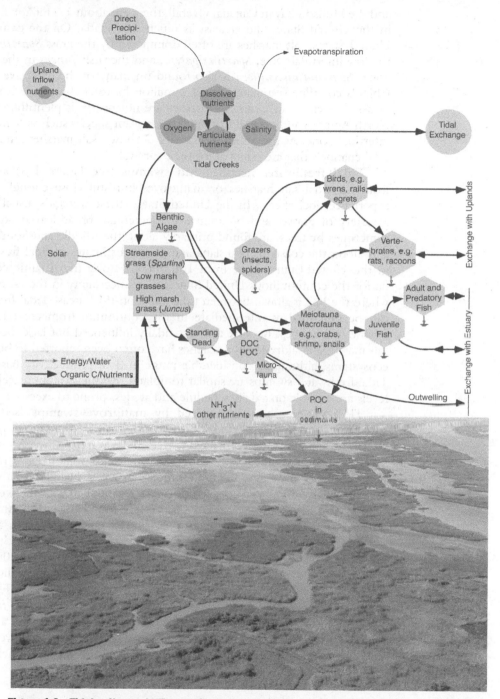

Figure 1-2 Tidal salt marsh: Energy flow diagram (top) and photograph of tidal salt marsh in Louisiana (bottom). Photo by W. J. Mitsch.

and the Hudson Bay in Canada. Overall, there are about 1.9 million ha of salt marshes in the United States and at least as much in Canada. On the eastern coast of the United States, salt marshes are often dominated by the grass *Spartina alterniflora* in the low intertidal zone, *Spartina patens*, and the rush *Juncus* in the upper intertidal zone. *Spartina alterniflora* is also found on many of the extensive salt marshes on China's coastline (estimated to be 2 million ha total) but there it is considered an invasive species. European salt marshes are not nearly as plentiful as those found in North America and generally do not have a plant species such as *S. alterniflora* in the intertidal zone. As the photo in Figure 1-2 shows, salt marshes are crisscrossed with tidal channels that are essential for their survival.

Tidal freshwater marshes and swamps (see Figure 1-3) are found inland from the tidal salt marshes or mangroves but still close enough to the coast to experience tidal effects. In the United States, these wetlands, usually dominated by a variety of grasses and by annual and perennial broad-leaved aquatic plants but sometimes by trees, are found primarily along the Middle and South Atlantic coasts and along the coasts of Louisiana and Texas. Estimates of tidal freshwater wetlands in the United States range from 160,000 ha along the Atlantic coast to 820,000 ha for the conterminous United States. The uncertainty in the estimate depends on where the line is drawn between tidal and non-tidal areas. Tidal freshwater marshes can be described as intermediate in the continuum from coastal salt marshes to freshwater marshes. Because they are tidally influenced but lack the salinity stress of salt marshes, tidal freshwater marshes have often been reported to be very productive ecosystems, although a considerable range in their productivity has been measured. Tidal freshwater swamps are similar to upland riverine swamps, except that the water levels are variable on a daily schedule and are less prone to excessive changes.

Tidal salt marshes are replaced by **mangrove swamps** (see Figure 1-4) in subtropical and tropical regions of the world. The word *mangrove* refers to both the wetland itself and to the salt-tolerant trees that dominate those wetlands. Mangrove swamps are found all over the world in tropical and subtropical regions, generally between 25°N and 25°S, and are estimated to cover 24 million ha worldwide. They are particularly dominant in the Indo-West Pacific region of the world, where more than half of the species of mangrove trees are found. In North America, mangrove wetlands are found on both coastlines in Mexico, where there are an estimated 940,000 ha of mangrove swamps (Martinez et al., 2007); in the United States, they are limited primarily to the southern tip of Florida where 300,000 to 500,000 ha are found. The extent of mangroves in North America (1.3 million ha) is a small fraction of the 24 million ha of mangroves found worldwide. In Florida, mangrove wetlands are generally dominated by the red mangrove tree *(Rhizophora)* and the black mangrove tree *(Avicennia)* and export organic matter to the adjacent estuaries, as do salt marshes. Like salt marshes, the mangrove swamps require protection from the open ocean and occur in a wide range of salinity and tidal influence.

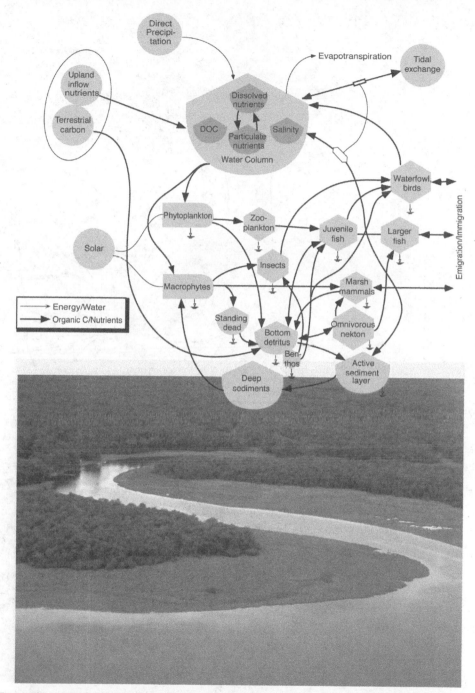

Figure 1-3 Tidal freshwater marsh: Energy flow diagram (top) and photograph of tidal fresh-water marsh in Maryland (bottom). Photo by A. H. Baldwin, reprinted with permission.

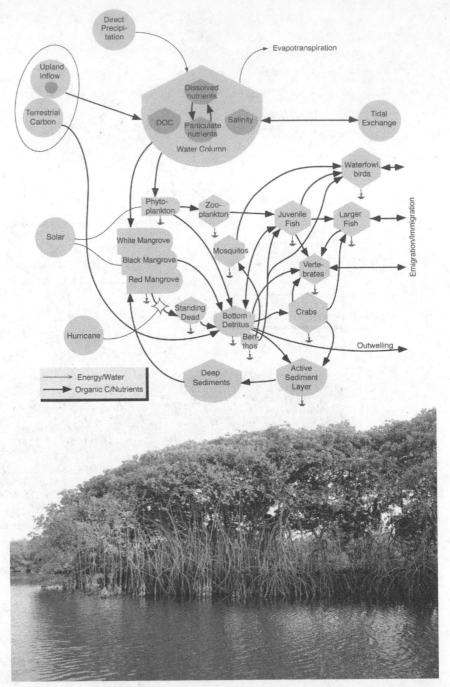

Figure 1-4   Mangrove swamp: Energy flow diagram (top) and photograph of mangrove swamp in Senegal, West Africa (bottom). Photo by W. J. Mitsch.

Freshwater Swamps and Marshes

On a real basis most of the wetlands of the world are not located along coastlines but are found inland (refer to Table 1-1). These wetlands are sometimes referred to as "non-tidal" in coastal regions to distinguish them from the coastal wetlands described previously. Our estimates in Table 1-1 suggest that 5.5 million km^2, or 95 percent of the total wetlands in the world, are inland. Frayer et al. (1983) estimated that in the lower 48 states in the United States, 32 million ha, or about 80 percent of the total wetlands in the lower 48 states, are inland. Including Alaska, there are a total of 107 million hectares of inland wetlands in the United States, representing 97% of the country's wetlands. It is difficult to divide these inland wetlands into simple categories. We have chosen to describe freshwater marshes and forested swamps in Chapter 3 and peatlands in Chapter 4.

Freshwater marshes (see Figure 1-5) includes a diverse group of wetlands characterized by emergent soft-stemmed aquatic plants such as cattail, bulrush, arrowhead, pickerel-weed, reed, and several other species of grasses and sedges, a shallow, seasonally changing water regime, and shallow organic soil deposits. These wetlands are ubiquitous and are estimated to cover 95 million ha around the world and about 27 million ha in the United States. Major regions where marshes dominate include the Okavango Delta in Botswana (see Figure 1-5 photo), the prairie pothole region of the Dakotas, and the Everglades of Florida. They occur in isolated basins, as fringes around lakes, and along sluggish streams and rivers.

Freshwater forested swamps range from wetlands that have standing water for most if not all of the growing season (see Figure 1-6; sometimes referred to as deepwater swamps) to riparian bottomland forests that are less frequently flooded but found all around the world across many climates (see Figure 1-7). The frequently flooded forested swamps occur in a variety of nutrient and hydrologic conditions—as alder (*Alnus*) swamps in Europe, as kahikatea (*Dacrycarpus*) swamps in New Zealand, and as cypress (*Taxodium*) and gum/tupelo (*Nyssa*) swamps in the southeastern United States. In the northern parts of the lower 48 United States, red maple (*Acer rubrum*) swamps are common, although there are no trees quite as adapted to flooding anywhere in North America as are *Taxodium* and *Nyssa*. Extensive tracts of riparian wetlands, which occur along rivers and streams, are occasionally flooded by those bodies of water but are otherwise dry for varying portions of the growing season. Riparian forests and deepwater swamps combined constitute the most extensive class of wetlands in the United States, covering from 22 to 25 million ha (Dahl and Johnson, 1991; Hall et al., 1994). In the eastern United States, riparian ecosystems, often referred to as bottomland hardwood forests, contain diverse vegetation that varies along gradients of flooding frequency. Riparian wetlands also occur in arid and semi-arid regions, where they are often a conspicuous feature of the landscape in contrast with the surrounding arid grasslands and desert. Riparian ecosystems are generally considered to be more productive than the adjacent uplands because of the periodic inflow of nutrients, especially when flooding is seasonal rather than continuous.

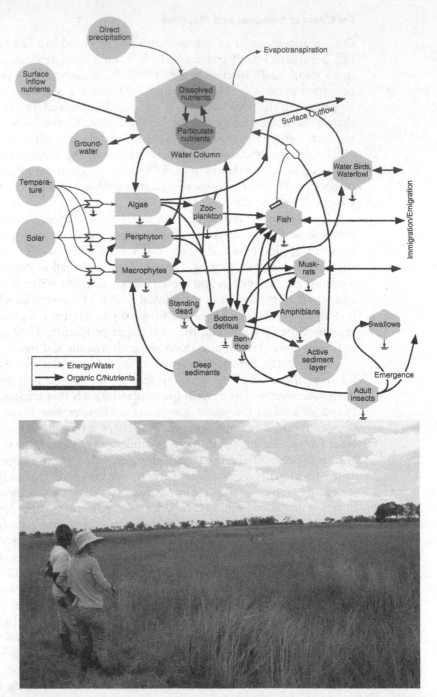

Figure 1-5   Freshwater marsh: Energy flow diagram (top) and photograph of freshwater marsh in Okavango Delta floodplain, Botswana, southern Africa (bottom). Photo by W. J. Mitsch.

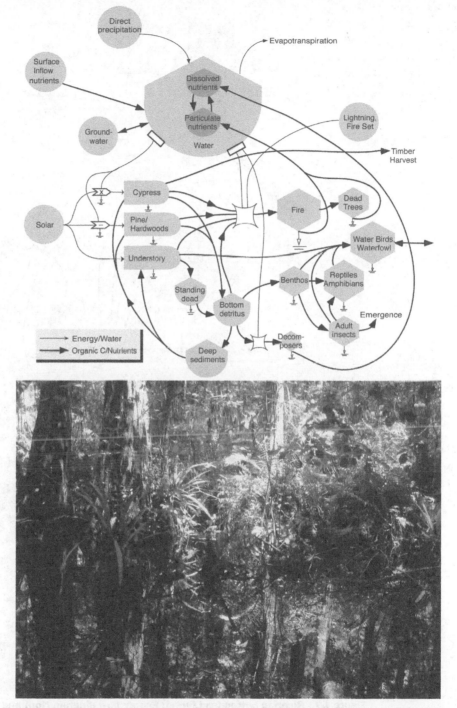

Figure 1-6 Freshwater swamp: Energy flow diagram (top) and photograph of freshwater swamp at Corkscrew Swamp, southwestern Florida (bottom). Photo by W. J. Mitsch.

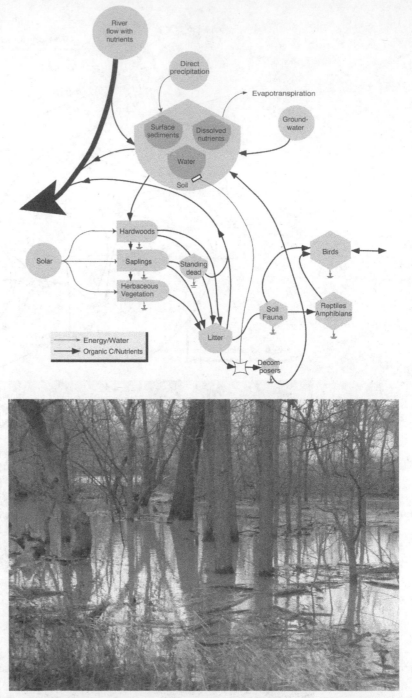

Figure 1-7 Riparian bottomland forest: Energy flow diagram (top) and photograph of riparian forested wetland in central Ohio (bottom). Photo by W. J. Mitsch.

Peatlands

As defined here, peatlands include the deep peat deposits of the boreal regions of the world (see Figure 1-8). They are the most ubiquitous wetland in the world, covering from 2.4 to 4.1 million km^2. A recent estimate of 3.3 million km^2 of peatlands was given by Wieder et al. (2006) for most of the world (where peatland coverage has been measured). In North America, the extensive peatlands of Alaska and Canada cover an estimated 0.52 and 1.11 million km^2, respectively (Zoltai, 1988; Hall et al., 1994), or 1.6 million km^2, a significant portion of the world's peatlands. Europe has about 0.96 million km^2 of peatlands, covering about 20% of the continent (Mitsch and Gosselink, 2007). The Western Siberian Lowland in Asia has about 0.79 million km^2 of peatlands. In total that is 3.37 million km^2. In the conterminous United States, peatlands are limited primarily to Wisconsin, Michigan, Minnesota, and the glaciated Northeast. Minnesota, with an estimated 2.7 million ha, has the largest peatland area in the United States (Glaser, 1987). There are also mountaintop bogs in the Appalachian Mountains of West Virginia such as those found in the Canaan Valley and peat-dominated wetlands, called pocosins, in the Coastal Plain of southeastern United States. Bogs and fens, the two major types of peatlands, occur as thick peat deposits in old lake basins or as blankets across the landscape. Many of these lake basins were formed by the last glaciation, and the peatlands are considered to be a late stage of a "filling-in" process. There is a wealth of European scientific literature on this wetland type, much of which has influenced the more recent North American literature on the subject. Bogs are noted for their nutrient deficiency and waterlogged conditions and for the biological adaptations to these conditions such as carnivorous plants and nutrient conservation.

Wetland Ecosystems Services

Wetlands are far more important in the biosphere than their 5 to 7% of the landscape suggests. They provide an immense storage of carbon that, if released with climate shifts, could accelerate those changes. They are known for their role in protecting clean water, so much so that thousands of wetlands have been constructed to clean all types of wastewater around the world. They protect coastlines from hurricanes and tsunamis, mitigate flooding of streams and rivers, and, most importantly to some, provide a bountiful habitat for a great diversity of plant and animal species. These "ecosystem services" of wetlands are discussed in great detail in Mitsch and Gosselink (2007) and are only summarized here.

Climate Stability

Wetlands as ecosystems represent one of the important linchpins of climate change. The amount of carbon stored in wetlands, particularly in boreal peatlands, is enormous (on the order of 30% of all the organic carbon storage in the planet) and any climate change that could affect that storage, say by drying the boreal regions, could result in a massive positive feedback of even more emissions of carbon dioxide to the atmosphere (see Chapter 10, Mitsch and Gosselink, 2007). Just as important, wetlands, especially

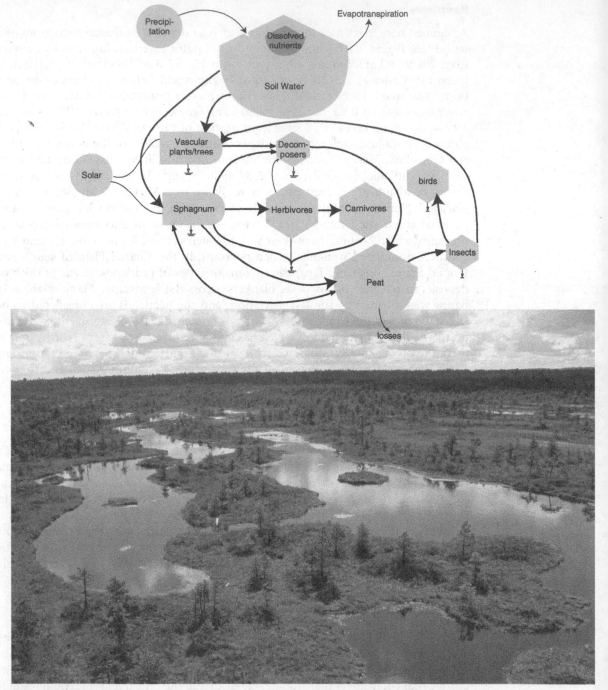

Figure 1-8 Peatland: Energy flow diagram (top) and photograph of peatland in Estonia (bottom). Photo by Valdo Kuusemets, reprinted with permission.

when they are in the early stages of succession or recovering from alterations such as beaver damming or herbivory, can be one of the best ecosystems on the planet for sequestering carbon from the atmosphere and permanently storing that carbon, first in plants, then in detrital matter, and finally as part of the soil structure itself.

On the other side of the ledger, wetlands are known sources of at least two so-called greenhouse gases—methane and nitrous oxides—and have been so for all the time that ecosystems have been on planet Earth. Methane production occurs when other anaerobic metabolisms have spent themselves and the soils are in extremely reduced conditions. Rates of carbon emission as methane are generally less than 10% of the rate at which carbon is being sequestered by the same wetland, but methane as a gas is 22 times more effective than carbon dioxide at adsorbing radiation in the atmosphere. While there have been many studies investigating methane emissions from wetlands, particularly in boreal regions, it has been extraordinarily difficult to extrapolate these studies to the biosphere level because of enormous landscape and hydrologic variability of wetland ecosystems.

Nitrous oxide (N_2O) is a powerful greenhouse gas, and it is often released in trace amounts by wetlands. That release is a byproduct of a desirable process in many wetlands—denitrification. Denitrification is the microbial reduction of nitrate-nitrogen under anaerobic conditions and its subsequent gaseous emissions. While most of the gaseous emissions from denitrification result in harmless nitrogen gas (N_2), a low percentage of the emissions occur as N_2O. Much more needs to be known about the balance between carbon sequestration in wetlands on the one hand and emission of trace greenhouse gases on the other.

Water Quality Improvement

Wetlands are sometimes referred to as *nature's kidneys* for the role that they actually and potentially play in improving water quality. It is indeed impressive to see how well wetlands with flow-through characteristics change water chemistry. They generally increase water clarity and remove chemicals such as nitrate-nitrogen. But they can be sources of organic matter, particularly if water entering the wetlands is low in organic matter, and often they can serve as sources of phosphorus if overloaded for a number of years. There are thousands of published papers in the past 35 years that have illustrated how natural and constructed wetlands improve water quality.

Coastal Protection

Coastal wetlands have been shown time and time again to be important coastline protection ecosystems. Major disasters just in the first decade of the 21st century have supported the cause for protecting coastal wetlands to protect humans and their resources, and conversely showing extraordinary damage when wetlands are removed. In December 2004, the great Indian Ocean tsunami caused by an earthquake off the coast of Indonesia led to 230,000 deaths around the entire Indian Ocean and billions of dollars of damage. Subsequent studies (e.g., Danielsen et al., 2005) showed that where coastal mangrove swamps were left in place to "bear the brunt" of the sometimes 10-m high tsunami waves, areas behind the mangroves were most protected. Conversely areas where mangrove swamps were drained or destroyed for commercial operations such as shrimp farms were devastated.

Eight months later, in late August 2005, the city of New Orleans was struck by the powerful Hurricane Katrina, causing billions of dollars of damage. Tens of thousands of people moved to higher grounds as a 6-m high storm surge overwhelmed New Orleans's levee system. New Orleans' real protection shield—the salt marshes of the Mississippi River delta—had been slowly eroding away for decades and the disaster was, unfortunately, predictable (Day et al., 2007) and may even happen again unless the salt marshes can be restored (Costanza et al., 2006).

Flood Mitigation

In addition to protecting coastal regions from storms, hurricanes, typhoons, and tsunamis, wetlands adjacent to rivers or in upstream reaches of watersheds are key to mitigating downstream river flooding. For example, with predictable and even more frequent occurrence, economically disastrous floods have occurred at least twice in a 15-year period (1993 and 2008) on the Upper Mississippi River Basin in the United States, partially as a result of river restrictions and development of floodplains into uses that are incompatible with flooding. Each time, there was an initial cry for restoring hundreds of thousands if not more wetlands in the region to be there to sponge the excess river water; but each time, before the water level even reached its base, the opportunity for flood mitigation with wetlands was forgotten and the basins were back to usual business.

Wildlife Protection

Wildlife often represents the entry point for the general public and wetlands. The wildlife in wetlands is fascinating and diverse. Wetlands are also known as *nature's supermarket* for the role that they play in supporting food chains, both aquatic and terrestrial. Wetlands are where critters go to eat or be eaten. Most people are familiar with birdbaths that are used in backyards to attract birds. Wetlands are *nature's birdbaths* in the sense that they attract sometimes hundreds of species, not for the water itself, but as a location of food, protection, and/or procreation. The vast Pantanal seasonal wetland in central South America is estimated to be the habitat for almost 500 species of birds. The Okavango Delta in south central Africa provides a pulsing wetland landscape that supports more than 400 bird species. If the wetland is connected to a stream, river, lake or ocean, wetlands can serve as fish nursery and feeding grounds. Isolated wetlands are important habitats for reptiles and amphibians, some of which depend on the wetland not being wet during the dry season. The value of wetlands for waterfowl, fish, and nature in general has always been known to naturalists, fishers, and hunters but not always by others; it is now known to school children everywhere through education programs that find wetlands to be good teaching opportunities.

Role of This Book

This book is meant as a stand-alone text that provides an ecosystem description (biotic and abiotic) of the major types of wetlands that are found in the world, simply categorized in Chapters 2, 3, and 4—coastal wetlands, freshwater marshes and forested swamps, and peatlands. The book provides descriptions of the ecosystem structure and function of these wetlands. Chapter 5 provides a review of the three fundamental ways that we can study wetland ecosystems in a "systems" way—mesocosms, full-scale experimental ecosystems, and mathematical modeling. As such, the text is a good "undergraduate" introduction to wetland ecosystems, especially when the course includes field trips to local wetlands. It is also meant to complement the textbook Wetlands, 4th edition (Mitsch and Gosselink, 2007) in upper-level undergraduate or graduate wetland courses. That book describes the human history and definitions of wetlands, and presents basic wetland science and applied wetland management, topics that are lightly touched upon in this book.

Recommended Readings

Mitsch, W. J., and J. G. Gosselink. 2007. *Wetlands, 4th ed.*, John Wiley & Sons, Hoboken, NJ.

Chapter 2

Coastal Wetlands

Coastal wetlands around the world can be divided into three general types: salt marshes, tidal freshwater wetlands, and mangrove swamps. The salt marsh, distributed worldwide along coastlines in middle and high latitudes, flourishes wherever the accumulation of sediments is equal to or greater than the rate of land subsidence and where there is adequate protection from destructive waves and storms. The important physical and chemical variables that determine the structure and function of the salt marsh include tidal flooding frequency and duration, soil salinity, soil permeability, and nutrient limitation, particularly by nitrogen. The vegetation of the salt marsh, primarily salt-tolerant grasses and rushes, develops in identifiable zones in response to these and possibly other factors. Mud and epiphytic algae are also often an important component of the autotrophic community. Heterotrophic communities are dominated by detrital food chains, with the grazing food chain being much less significant except in recent marsh die-off episodes.

Freshwater coastal wetlands include both marshes and swamps. Freshwater tidal marshes combine many features of both salt marshes and inland marshes. They act in many ways like salt marshes, but the biota reflect the increased diversity made possible by the reduction of the salt stress. Plant diversity is high, and more birds use these marshes than any other marsh type. Because they are inland from the saline parts of the estuary, they are often close to urban centers. This makes them more prone to human impact than coastal salt marshes. Along coastal rivers, tidal freshwater swamps tend to occupy a narrow range at the furthest extent of the tidal range. These forests are most prevalent along larger rivers with flat topography and considerable discharge. They occur where tidal waters are normally fresh and shallow enough for tree establishment.

Mangrove swamps replace salt marshes as the dominant coastal ecosystems in subtropical and tropical regions. An estimated 240,000 km^2 of mangrove *wetlands are found throughout the world. Mangrove wetlands are limited in the United States to the southern extremes of Florida (where there are approximately 5,000 km^2 of mangroves) and to emerging mangrove swamps on the Louisiana coastline. Mangrove wetlands have been classified according to their hydrodynamics and topography as fringe mangroves, riverine mangroves, basin mangroves, and dwarf or scrub mangroves. The dominant mangrove plant species are known for several adaptations to the saline wetland environment, including prop roots, pneumatophores, salt exclusion, salt excretion, and the production of viviparous seedlings. Their productivity and organic export are closely related to their hydrogeomorphic setting.*

Several types of wetlands in coastal areas are influenced by alternating floods and ebbs of tides. Coastal wetlands include tidal salt marshes, tidal freshwater wetlands (marshes and forests), and mangrove swamps (see Figure 2-1). Near coastlines, the salinity of the water approaches that of the ocean (35 ppt = parts per thousand), whereas further inland, the tidal effect can remain significant even when the salinity

Figure 2-1 Coastal wetlands: a. tidal salt marsh in New Jersey; b. tidal freshwater marsh in Maryland; c. mangrove swamp in Florida. Photo b courtesy of A. Baldwin, reprinted with permission; photo c by W. J. Mitsch.

is that of fresh water. Tidal freshwater wetlands are found upstream of salt water (0.5 ppt = 500 ppm (parts per million) and lower salinity), while salt marshes and mangroves are found downstream (salt marshes in temperate and boreal zones; mangroves in tropics) in polyhaline and mesohaline estuarine waters greater than 5 ppt (= 5,000 ppm) in salinity (see Figure 2-2). These coastal wetlands are often found in abundance in the river deltas and estuaries of the world—where large rivers debouch onto low-energy coasts (see Figure 2-3). These river deltas and estuaries span the world's latitudes and climatic zones. In the tropics, tidally influenced wetlands of these deltas are mangroves. Above 25° latitude, mangroves give way to salt marshes. In North America, large deltas are restricted to the coasts of the South Atlantic and the Gulf of Mexico. The Mississippi River deltaic marshes are the major example of this type of development and support the most extensive coastal marshes in the United States.

It is not known with accuracy how many coastal wetlands there are in the world, but it is a small fraction of the 6 to 10 million km^2 of wetlands estimated for the entire earth. About 0.24 million km^2 of mangroves alone are estimated to be found worldwide and tidal marshes (freshwater and salt) probably cover a similar amount of area. The total area of wetlands considered coastal or estuarine wetlands in the United

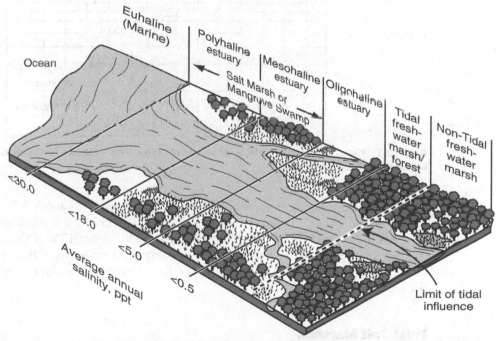

Figure 2-2 Coastal wetlands lie on gradients of decreasing salinity from inland to the ocean. Where salinity is sufficient, salt marshes (in temperate zone) and mangroves (in tropics) are found. Tidal freshwater marshes and forests still experience tides but are above the salt boundary. Further inland are marshes and forested swamps that experience neither salt nor tides.

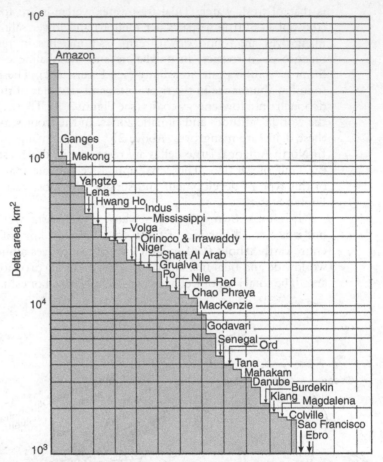

Figure 2-3 The area of deltaic plains of selected major river systems of the world. (*After Coleman and Roberts, 1989*)

States, including Alaska, is approximately 3.2 million ha, with about 1.9 million ha as salt marsh and 0.5 million ha as mangrove (see Table 2-1). Alaskan estuarine wetlands are estimated to cover 0.86 million ha, with about 17% of these wetlands (0.15 million ha) vegetated and thus presumably salt marsh. More than half of the salt marshes (53%) in the United States are found in the Mississippi River Delta to the Gulf of Mexico.

Tidal Salt Marshes

Salt marshes are found throughout the world along protected coastlines in the middle and high latitudes (see Figure 2-4a). Salt marshes can be narrow fringes on steep shorelines or expanses that are several kilometers wide. They are found near river

Table 2-1 Estimated Area of Coastal Wetlands in the United States (× 1,000 ha)

	Tidal Salt Marsh[a]	Tidal Freshwater Marsh[b]	Mangrove[b]	Total
Atlantic Coast	669	400	-	1,069
Gulf of Mexico	1,011	362	506	1,879
Pacific Coast	49	57	-	106
Alaska[c]	146	-	-	146
Total	1,875	819	506	3,200

[a]Watzin and Gosselink (1992).
[b]Field et al. (1991). Mangroves are estimated as forested wetlands in Ten Thousand Islands Drainage Area, southwest Florida.
[c]Hall et al. (1994).

mouths, in bays, on protected coastal plains, and around protected lagoons. Different plant associations dominate different coastlines, but the ecological structure and function of salt marshes is similar around the world. Salt marshes, dominated by rooted vegetation that is alternately inundated and dewatered by the rise and fall of the tide, appear from afar to be vast fields of grass of a single species. In reality, salt marshes have a complex zonation and structure of plants, animals, and microbes, all tuned to the stresses of salinity fluctuations, alternate drying and submergence, and extreme daily and seasonal temperature variations. A maze of tidal creeks with plankton, fish, nutrients, and fluctuating water levels crisscrosses the marsh, forming conduits for energy and material exchange with the adjacent estuary. Studies of a number of different salt marshes have found them to be highly productive and to support the spawning and feeding habits of many marine organisms. Thus, salt marshes and tropical mangrove swamps throughout the world form an important interface between terrestrial and marine habitats.

Geographic Extent

Salt marshes are found near river mouths, in bays, on protected coastal plains, and around protected lagoons. Different plant associations dominate different coastlines, but the ecological structure and function of salt marshes are similar around the world. Based on the classification system developed by Valentine Chapman (1960, 1976a), the world's salt marshes can be divided into the following major geographical groups:

- **Arctic:** This group includes marshes of northern Canada, Alaska, Greenland, Iceland, northern Scandinavia, and Russia. Probably the largest extent of marshes in North America, as much as 300,000 km^2, occurs along the southern shore of the Hudson Bay. These marshes, influenced by ice, extreme low temperatures, a positive water balance, and numerous inflowing streams, can be generally characterized as brackish rather than saline. Various species of the sedge *Carex* and the grass *Puccinellia phryganodes* often dominate. Parts

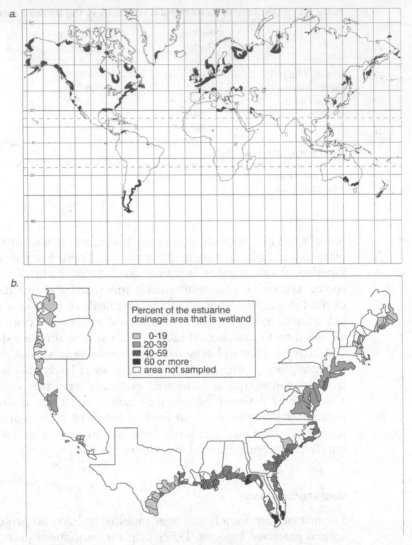

Figure 2-4   Distribution of a. salt marshes of the world and b. wetlands in coastal drainage areas of the United States, including freshwater tidal wetlands and mangroves as well as tidal salt marshes. (*After Chapman, 1977 and Field et al., 1991*)

of the southwestern coast of Alaska are dominated by species of *Salicornia* and *Suaeda*.

- **Northern Europe:** This group includes marshes along the west coast of Europe from the Iberian Peninsula to Scandinavia, including Great Britain and the Baltic Sea coast. Most of the western European coastal environment is characterized by a moderate climate with sufficient precipitation but high salinities toward the southern extremes. Dominant species include *Puccinellia maritima, Juncus gerardi, Salicornia* spp., and *Spartina anglica* and *S. townsendii*. The west coast of Great Britain and parts of the Scandinavian and Baltic Sea coasts, where substrates are dominated by sand and salinities are low, are populated by *Festuca rubra, Agrostis stolonifera, Carex paleacea, Juncus bufonius, Desmoschoenus bottanica*, and *Scripus* spp. The muddy coast of the English Channel is dominated by *Spartina townsendii*. Salt marshes in northern Europe are often characterized by a lack of vegetation in the intertidal zone, in contrast to North American marshes.

- **Mediterranean:** This group includes the arid, rocky-to-sandy, high-salinity coasts of the Mediterranean Sea. The salt marshes are dominated by low shrubby vegetation, *Arthrocnemum, Limonium, Juncus* spp., and the halophyte *Salicornia* spp.

- **Eastern North America:** These marshes, mostly dominated by *Spartina* and *Juncus* species, are found along the eastern coasts of the United States and Canada and the Gulf Coast of the United States. Salt marshes are most prevalent along the eastern coast of the United States from Maine to Florida and on into Louisiana and Texas along the Gulf of Mexico (see Figure 2-4b). The Eastern North American group is further divided into three subgroups:

  - *Bay of Fundy:* River and tidal erosion is high in the soft rocks of this region, producing an abundance of reddish silt. The tidal range, as exemplified at the Bay of Fundy, is large, leading to a few marshes in protected areas and considerable depth of deposited sediments. *Puccinellia americana* dominates the lower marsh, and *Juncus balticus* is found on the highest levels.

  - *New England:* Marshes are built mainly on marine sediments and marsh peat, and there is little transport of sediment from the hard-rock uplands. These marshes range from Maine to New Jersey and are dominated by *Spartina alterniflora* in the low marsh, with *S. patens* mixed with *Distichlis spicata* in the high marsh.

  - *Coastal Plain:* These marshes extend southward from New Jersey along the southeastern coast of the United States to Texas along the Gulf of Mexico. Major rivers supply an abundance of silt to the recently elevated Coastal Plain. The tidal range is relatively small. The marshes are laced with tidal creeks. Mangrove swamps replace salt marshes along the southern tip of Florida. Because of the extensive delta marshes built by the Mississippi River, the Gulf Coast contains about 60% of the coastal salt and fresh marshes of the United States (see Figure 2-4b). Dominant species are *Spartina alterniflora, S. patens, Juncus roemerianus*, and *Distichlis spicata*.

- **Western North America:** Compared with the Arctic and the eastern coast of North America, salt marshes are far less abundant along the western coasts of the United States and Canada because of the geomorphology of the coastline. On this rugged coast with its Mediterranean-type climate is found a narrow belt of *Spartina foliosa*, often bordered by broad belts of *Salicornia* and

Suaeda. Spartina alterniflora is a non-native invasive plant in coastal marshes north of California.

- **Australasia:** Salt marshes are frequently found in river deltas along the temperate coastlines of eastern Asia, Australia, and New Zealand on the Pacific Ocean, Indian Ocean, and Tasman Sea.

  - *Eastern Asia:* The coasts of China, Japan, Russia, and Korea are generally rugged and uplifted, with moderate precipitation but limited marsh development. *Triglochin maritima, Limonium japonicum, Salicornia*, and *Zoysia macrostachya* dominate these marshes. Major areas of salt marsh restoration have occurred on China's eastern coastline because of the introduction of *Spartina anglica* and *S. alterniflora*.

  - *Australia:* This group also includes New Zealand and Tasmania. It is characterized by high rainfall and geographic isolation. Cosmopolitan species in Australian salt marshes include *Sporobolus virginicus, Sarcocornia quinqueflora*, and *Suaeda australis*. However, invasion of salt marshes of Australia and New Zealand by several species is common. In New Zealand and other temperate-zone salt marshes of the region, *Spartina anglica* is a major invasive species. Even with less rainfall and a clearly defined seasonal pattern of wet and dry on the western coast of Australia, salt marshes can be found, particularly around Shark Bay and the Peel–Harvey estuaries. Unlike the general case around much of the world, a majority of the salt marshes of Australia are found in tropical regions (Adam, 1998).

- **South America:** South American coasts too far south and too cold for mangroves are rugged and geographically isolated. They are dominated by unique species of *Spartina, Limonium, Distichlis, Juncus, Heterostachys*, and *Allenrolfea*.

- **Tropics:** Although mangroves generally dominate tropical coastlines, salt marshes are found in the tropics on high-salinity flats that mangroves cannot tolerate. *Spartina* spp. and the halophytic genera *Salicornia* and *Limonium* often dominate.

Hydrogeomorphology

The physical features of tides, sediments, freshwater inputs, and shoreline structure determine the development and extent of salt marsh wetlands within their geographical range. Coastal salt marshes are predominantly intertidal; that is, they are found in areas at least occasionally inundated by high tide but not flooded during low tide. A gentle, rather than steep, shoreline slope allows for tidal flooding and the stability of the vegetation. Adequate protection from wave and storm energy is also a physical requirement for the development of salt marshes. Sediments that build salt marshes originate from upland runoff, marine reworking of the coastal shelf sediments, or organic production within the marsh itself.

Hydrology

Tidal energy represents a subsidy to the salt marsh that influences a wide range of physiographic, chemical, and biological processes, including sediment deposition and scouring, mineral and organic influx and efflux, flushing of toxins, and the control of sediment redox potential. These physical factors in turn influence the species that occur on the marsh and their productivity. The tide range generally sets the lower and upper limits of the marsh. The lower limit is set by the depth and the duration of flooding and by the mechanical effects of waves, sediment availability, and erosional forces. The upland side of the salt marsh generally extends to the limit of flooding on extreme tides, normally between mean high water and extreme high water of spring tides. Based on marsh elevation and flooding characteristics, the marsh is often divided into two zones, the upper marsh (*high marsh*) and the intertidal lower marsh (*low marsh*) (see Table 2-2). The high marsh is flooded irregularly and can experience at least 10 days of continuous exposure to the atmosphere, whereas the low marsh is flooded almost daily, and there are never more than nine continuous days of exposure. In the Gulf Coast marshes of the United States, the terms *streamside marshes* and *inland marshes* generally replace low and high marsh, respectively, because in these flat, expansive marshes the streamside levees are actually the highest marsh elevations.

Marsh Development

Although a number of different patterns of development can be identified, salt marshes can be classified broadly into 1) those that were formed from reworked marine sediments on marine-dominated coasts, and 2) those that were formed in deltaic areas where the main source of mineral sediment is riverine.

Marine-dominated marshes are typical of most of the world's coastlines. On marine-dominated coasts, salt marsh development requires sufficient shelter to ensure sedimentation and to prevent excessive erosion from wave action. Marshes can develop at the mouths of estuaries where sediments are deposited by the river, behind spits and bars, and in bays that offer protection from waves and long-shore currents. A spit is a neck of land that acts to trap sediment on its lee side and protects the marsh from the full forces of the open sea. The most extensive examples of this type of coastal

Table 2-2 Hydrologic Demarcation between Low Marsh and High Marsh in Salt Marshes

| Marsh | Submergences | | Maximum Period of Continuous Exposure (days) |
	Per Day in Daylight	Per Year	
High marsh	<1	<360	≥10
Low marsh	>1.2	>360	≤9

Source: Adapted from Chapman (1960).

salt marsh in the United States have developed behind outer barrier reefs along the Georgia-Carolina coast. Several large bays, such as Chesapeake Bay, Hudson Bay, the Bay of Fundy, and San Francisco Bay, are also adequately protected from storms and waves so that they can support extensive salt marshes. These salt marshes in bays have features of both marine and deltaic origins. They occur on the shores of estuaries where shallow water and low gradients lead to river sediment deposition in areas protected from destructive wave action. Tidal action must be strong enough to maintain salinities above about 5 ppt; otherwise, the salt marsh will be replaced by reeds, rushes, and other freshwater aquatic plants.

Tidal Creeks

A notable physiographic feature of salt marshes, especially low marshes, is the development of *tidal creeks* in the marsh itself (see Figure 2-5). These creeks develop, as do rivers, "with minor irregularities sooner or later causing the water to be deflected into definite channels" (Chapman, 1960). The creeks serve as important conduits for material and energy transfer between the marsh and its adjacent body of water. A tidal creek has salinity similar to that of the adjacent estuary or bay, and its water depth varies with tide fluctuations. Its microenvironments include different vegetation zones

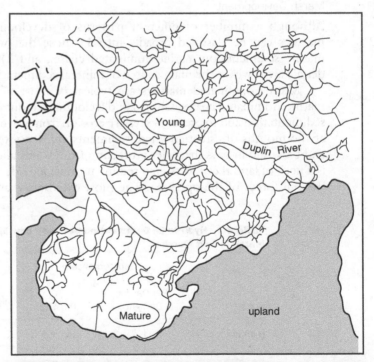

Figure 2-5   Drainage patterns of tidal creeks in young and mature *Spartina alterniflora* salt marshes in the Duplin River drainage, Doboy Sound, Georgia. (*After Wiegert and Freeman, 1990, and Wadsworth, 1979*)

along its banks that have aquatic food chains important to the adjacent estuaries. Because the flow in tidal channels is bidirectional, the channels tend to remain fairly stable; that is, they do not meander as much as streams that are subject to a unidirectional flow. As marshes mature and sediment deposition increases elevation, however, tidal creeks tend to fill in and their density decreases (see Figure 2-5).

Pannes

A distinctive feature of salt marshes is the occurrence of pannes (pans). The term *panne* is used to describe bare, exposed, or water-filled depressions in the marsh, which may have different sources. In the higher reaches of the marsh, inundated by only the highest tides, *sand barrens* appear where evaporation concentrates salts in the substrate, killing the rooted vegetation. Thin films of blue-green algae often cover these exposed barrens. *Mud barrens* are naturally occurring depressions in the marsh that are intertidal and retain water even during low tide. Pannes are often devoid of vascular vegetation or support submerged or floating vegetation because of the continuous standing water and the elevated salinities when evaporation is high and are continually forming and filling due to shifting sediments and organic production. The vegetation that develops in a mud panne, for example, wigeon grass (*Ruppia* sp.), is tolerant of salt at high concentrations in the soil water. Relatively permanent ponds are formed on some high marshes and are infrequently flooded by tides. Because of their shallow depth and their support of submerged vegetation, they are used heavily by migratory waterfowl. Pannes are a common feature due to human intervention, occurring where free tidal movement has been blocked by roads or levees, where spoil deposits have elevated a site, or where soil excavation, for example, for highway construction, has occurred in a marsh.

Soils and Salinity

The sediment source and tidal current patterns determine the sediment characteristic of the marsh. Salt marsh sediments can come from river silt, organic productivity in the marsh itself, or reworked marine deposits. As a tidal creek rises out of its banks, water flowing over the marsh slows and drops its coarser-grained sediment load near the stream edge, creating a slightly elevated streamside levee. Finer sediments drop out farther inland, giving rise to the well-known "streamside" effect, characterized by the greater productivity of grasses along tidal channels than inland, a result of the slightly larger nutrient input, higher elevation, and better drainage.

Salt marshes that experience a large tide range (e.g., the Wash, England) tend to approximate the ambient marine water salinity even though rainfall may be significant. In coastal marshes adjacent to large rivers, on the other hand (e.g., the north coast of the Gulf of Mexico), fresh water dilutes marine sources and the marshes are brackish or even fresh. Extreme salinities can be found in subtropical areas such as the Texas Gulf Coast, where rivers and rainfall supply little fresh water and tides have a narrow range so that flushing is reduced. As a result, marine water is concentrated by evapotranspiration, often to double seawater strength or even higher.

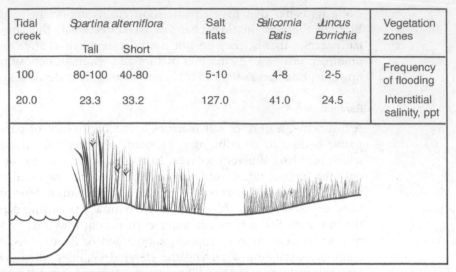

Tidal creek	Spartina alterniflora		Salt flats	Salicornia Batis	Juncus Borrichia	Vegetation zones
	Tall	Short				
100	80-100	40-80	5-10	4-8	2-5	Frequency of flooding
20.0	23.3	33.2	127.0	41.0	24.5	Interstitial salinity, ppt

Figure 2-6   The relation of a salt flat's interstitial soil salinity and its vegetation. (*After Wiegert and Freeman, 1990, and Antlfinger and Dunn, 1979*)

Lateral salinity gradients develop as a function of flooding frequency and subsequently influence vegetation productivity (see Figure 2-6). Near the adjacent tidal creek, frequent tidal inundation keeps sediment salinity at or below sea strength. As the marsh elevation increases, the inundation frequency decreases and the finer sediments drain poorly. At the elevation shown in Figure 2-6 as salt flats, infrequent spring tides bring in salt water that is concentrated by evaporation. Flushing is not frequent enough to remove these salts, so they accumulate to lethal levels. Above this elevation, tidal flooding is so infrequent that salt input is restricted, and flushing by rainwater is sufficient to prevent salt accumulation. In this way, the salt gradient set up by the interaction of marsh elevation, tides, and rain often controls the general zonation pattern of vegetation and its productivity. Within the salt marsh zone itself, however, all plants are salt tolerant, and it is misleading to account for plant zonation and productivity on the basis of salinity alone. Salinity, after all, is the net result of many hydrodynamic factors, including slope and elevation, tides, rainfall, freshwater inputs, and groundwater. Thus, when *Spartina* flourishes in the intertidal zone, it is also responding to tides that reduce the local salinity, remove toxic materials, supply nutrients, and modify soil anoxia. All of these factors collectively contribute to different productivities and different growth forms in the intertidal and high marshes.

Vegetation

The salt marsh ecosystem has diverse biological components, which include vegetation and animal and microbe communities in the marsh and plankton, invertebrates, and fish in the tidal creeks and estuaries. The discussion here will be limited to the biological structure of the marsh itself. Plants and animals in these systems have adapted to the stresses of salinity, periodic inundation, and extremes in temperature.

The vegetation of salt marshes can be divided into zones that are related to the high and low marshes described previously but that also reflect regional differences. Figure 2-7 shows a typical New England vegetation zonation pattern from streamside to upland. The intertidal zone or low marsh next to the estuary, bay, or tidal creek is dominated by the tall form of *S. alterniflora* Loisel (smooth cordgrass). In the high marsh, *S. alterniflora* gives way to extensive stands of *S. patens* (saltmeadow cordgrass) mixed with *Distichlis spicata* (spikegrass) and occasional patches of the shrub *Iva frutescens* (marsh elder) and other shrubs. Beyond the *S. patens* zone and at normal high tide, *Juncus gerardi* (blackgrass) forms pure stands. At the upper edge of a marsh inundated only by spring tides, two groups of species are common, depending on the local rainfall and temperature. Where rainfall exceeds evapotranspiration, salt-tolerant species give way to less tolerant species such as *Panicum virgatum* (switchgrass), *Phragmites australis* (common reed), *Limonium carolinianum* (sea lavender), *Aster* spp. (asters), and *Triglochin maritima* (arrow grass). On the southeastern New England coast where evapotranspiration may

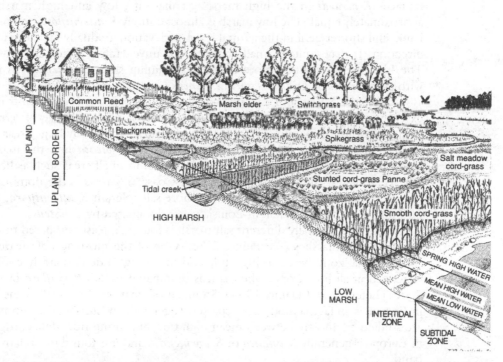

Figure 2-7   Idealized zonation of communities on a typical North Atlantic salt marsh. The location of the different plant associations is strongly influenced by small differences in elevation above the mean high water level. (*After Dreyer and Niering, 1995, reprinted by permission of Tom Ouellette, Vernon, CT*)

exceed rainfall during the summer, salts can accumulate in these upper marshes, and salt-tolerant halophytes such as *Salicornia* spp. (saltwort) and *Batis maritima* flourish. Bare areas with salt efflorescence are common. Other features of New England salt marshes include well-flushed mosquito ditches lined with tall *S. alterniflora* and salt pannes containing short-form *S. alterniflora*.

Crain et al. (2004) used greenhouse and field transplants along a New England coast to compare biotic and abiotic factors influencing salt marsh plants. Salt marsh plants transplanted into freshwater marsh conditions without competitors grew better than in the salt marsh. However, when salt marsh plants were transplanted into a freshwater marsh with neighboring plants, they were out competed by the freshwater marsh plants. The authors surmised that plants in environmental extremes such as salt marshes are determined by their tolerance to physiological stress while the occurrence of inland wetland plants is dictated by their competitive ability.

Characteristic patterns of vegetation found in other salt marshes are shown in Figure 2-8. South of the Chesapeake Bay along the Atlantic Coast, salt marshes typical of the Coastal Plains appear (see Figure 2-8a). These marshes are similar in zonation to those in New England except that (1) tall *S. alterniflora* often forms only in very narrow bands along creeks, (2) the short form of *S. alterniflora* occurs more commonly in the wide middle zone, and (3) *Juncus roemerianus* (black rush) replaces *J. gerardi* in the high marsh. At maturity, low and high marsh areas are approximately equal. The low marsh is almost entirely *S. alterniflora*, tall on the creek bank, and shorter behind the natural levee as elevation gradually increases in an inland direction. It may contain small vegetated or unvegetated ponds and mud barrens. The high marsh is much more diverse, containing short *S. alterniflora* intermixed with associations of *Distichlis spicata*, *Juncus roemerianus*, and *Salicornia* spp.

Along the Mississippi and northwest Florida coasts, *J. roemerianus* is found in extensive monocultures (see Figure 2-8b). There is often a fringe of *S. alterniflora* along the seaward margin, followed, in an inland direction, by large areas of tall and short *J. roemerianus*. Mixtures of *S. patens* and *D. spicata* line the marsh on the landward edge, and *Salicornia* spp. can be found in small areas such as berms where salt accumulates. Along the northern Gulf Coast, *S. patens* is the dominant species, occurring in a broad zone inland of the more salt-tolerant *S. alterniflora*. More than 200,000 ha of coastal marsh in Louisiana are dominated by *S. patens*.

In Europe, a totally different salt marsh is found, at least compared to the eastern United States marshes (see Figure 2-8c). One of the most notable features is that the intertidal zone between high tide and mean high tide is sparsely covered if it is vegetated at all in Europe, whereas it is dominated by *S. alterniflora* in the United States (Lefeuvre and Dame, 1994). So much of what would be called the low marsh in Europe is, in fact, a mud flat or sparsely vegetated. In Europe, the salt marshes that are studied are mostly between mean high tide and spring tide. The cordgrass found in Europe is generally *S. anglica* or *S. townsendii*, and it is found in a relatively narrow band.

In the last 15–20 years, the clonal grass *Elymus athericus* has spread into the middle and low marshes of many European salt marshes. This invasive plant, because

a. Southeast Atlantic coast salt marsh

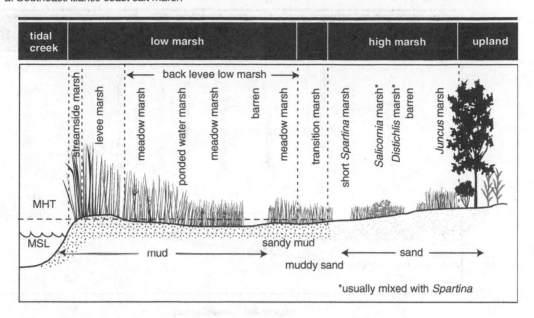

b. Eastern Gulf of Mexico coast salt marsh

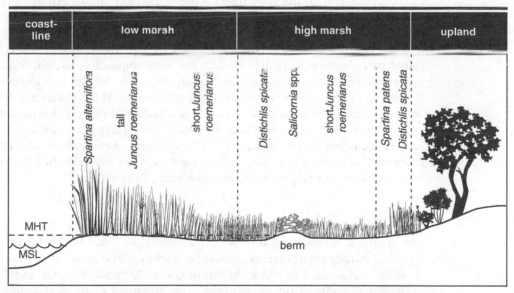

Figure 2-8 Zonation of vegetation in typical salt marshes: a. Southeastern USA Atlantic Coast, b. Eastern and Northern Gulf of Mexico, c. Northern France. (a. After Wiegert and Freeman, 1990; b. After Montague and Wiegert, 1990; c. After LeFeuvre and Dame, 1994)

c. European salt marsh

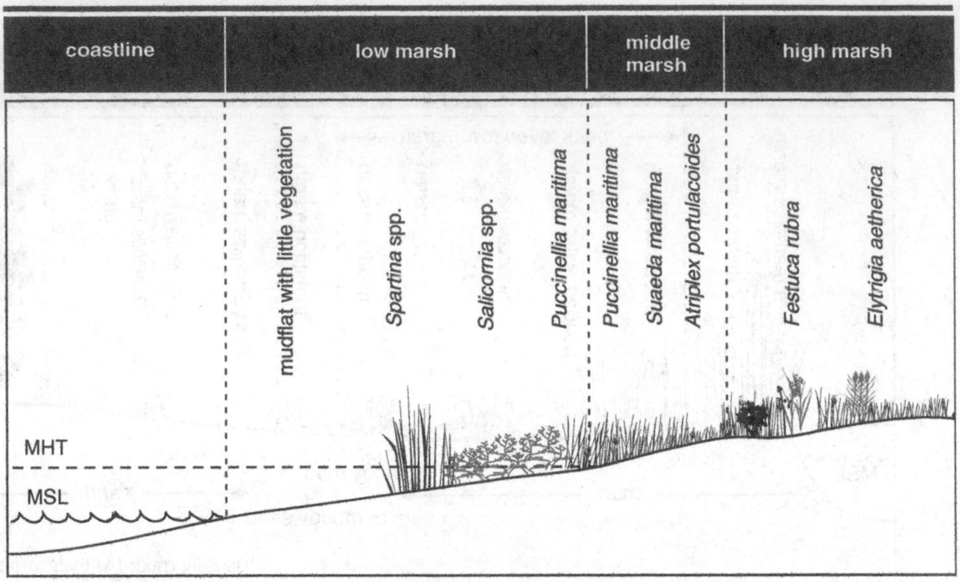

Figure 2-8 (*continued*)

of its large size, traps macrodetritus on the marsh, limiting its export to the adjacent estuary (Bouchard and Lefeuvre, 2000; Lefeuvre et al., 2003; Valéry et al., 2004), and has a significant impact on salt marsh biodiversity (Pétillon et al., 2005).

Salt marshes closer to the polar regions are less understood. Funk et al. (2004) found that elevation, conductivity, and soil ion composition all contributed to plant cover and species composition in an Alaskan salt marsh. As elevation increased, salinity decreased, resulting in increased plant species richness. At the lowest marsh elevations, only *Puccinella phryganodes* occurred. Mid-elevation sites were dominated by *Carex subspathaceae* and high-elevations sites had the highest cover with as many as 16 species, including *Dupontia fischeri* and *Eriophorum angustifolium*. Zhu et al. (2008) reported that biomass in coastal tundra marshes in eastern Antarctica were dominated by a combination of algae, moss, cyanobacteria, and bacteria.

Consumers

Salt marshes, whose features are characteristic of both terrestrial (aerobic) and aquatic (anoxic) environments, provide a harsh environment for consumers. Salt is an additional stress with which they must contend. In addition, the variability of the environment through time is extreme. The dominant plant food source for marsh consumers is generally a marsh grass, which is usually limited in its nutritional value. Considering all these limitations, the number of consumers in the salt marsh is surprisingly diverse.

Many faunal species, particularly vertebrate taxa, utilize tidal marshes (both salt and freshwater) as a component of a larger set of coastal ecosystems. There are, however, many reported species that are considered endemic to tidal marshes. In a global review of terrestrial vertebrates and their occurrence in tidal marshes, Greenberg et al. (2006) found 25 species (or subspecies) that were endemic to tidal marshes. Interestingly, nearly all of the species were restricted to North America. Sampling bias may explain some of this; however, very few records of tidal endemics were reported in well-studied regions such as Europe and Australia. Another possible factor is the large area of tidal marshes present in North America, and the higher occurrence of endemism is a reflection of a species-area relationship. While these factors or others may contribute to the high North American endemism, there is currently no comprehensive theory for this phenomenon.

It is convenient to classify consumers according to the type of marsh habitat they occupy, although the animals, especially in the higher trophic levels, move from one habitat to another. The marsh can be divided into three major habitats: an *aerial habitat*, the aboveground portion of the macrophytes, which is seldom flooded; a *benthic habitat*, the marsh surface and lower portions of the living plants; and an *aquatic habitat*, the marsh pools and creeks (see Figure 2-9).

Aerial Habitat

The aerial habitat is similar to a terrestrial environment and is dominated by insects and spiders that live in and on the plant leaves. This is the grazing portion of the

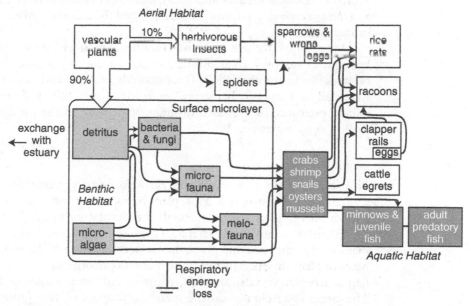

Figure 2-9 Salt marsh food web, showing the major producer and consumer groups of the aerial habitat, benthic habitat, and aquatic habitat. (*After Montague and Wiegert, 1990*)

salt marsh food web. The most common leaf-chewing organisms in salt marshes in the eastern United States are the arthropod *Orchelimum*, the weevil *Lissorhoptrus*, and the squareback crab *Sesarma*. In addition, there are abundant sap-sucking insects (*Prokelisia marginata, Delphacodes detecta*) that ingest material translocated through the plant's vascular tissue or empty the contents of mesophyll cells. Numerous carnivorous insects are also found in this habitat. Pfeiffer and Wiegert (1981) listed 81 species of spiders and insects in North Carolina, South Carolina, and Georgia *Spartina* marshes.

Salt marshes support large populations of wading birds, including egrets, herons, willets, and even woodstorks and Roseate Spoonbills. Coastal marshes also support vast populations of migratory waterfowl, including the Mallard (*Anas platyrhynchos*), American Wigeon (*A. americana*), Gadwall (*A. strepena*), Redheads (*Aythya americana*), and Teals (*Anas discors* and *A. crecca*). Black Duck (*A. rubripes*) is a permanent resident in many marshes, as are a number of songbirds.

A number of birds, including the Marsh Wren (*Cistothorus palustris*) and the Seaside Sparrow (*Ammodramus maritimus*), Laughing Gulls (*Larus atricilla*), and Forster's (*Sterna forsteri*) and Common Terns (*S. hirundo*), feed and nest in the marsh grasses. Wrens feed primarily on insects, and the sparrows apparently feed on the marsh surface, eating worms, shrimp, small crabs, grasshoppers, flies, and spiders. The Clapper Rail (*Rallus longirostris*) is another permanent marsh resident, feeding primarily on cutworm moths and small crabs. Many nonpermanent insectivorous birds forage in the salt marsh periodically, entering from adjacent fresher marshes, beaches, and upland habitats or migrating through. These include the Sharptailed Sparrow (*Ammodramus caudacutus*), swallows (*Tachycineta bicolor, Hirundo rustica*, and *Stelgidopteryx serripennis*), Red-winged Blackbirds (*Agelaius phoeniceus*), and various gulls.

Migratory waterfowl use coastal marshes extensively, mostly as wintering grounds, but also as stopover areas during fall and spring migrations. In some areas, geese or duck flocks numbering in the hundreds of thousands denude coastal marshes. Repeated and intense herbivory, especially when followed by high water levels, salinity extremes, or extended drought, may result in the formation of mud flats or shallow open-water ponds.

Benthic Habitat

Probably less than 10% of the aboveground primary production of the salt marsh is grazed by aerial consumers. Most plant biomass dies and decays on the marsh surface, and its energy is processed through the detrital pathway. The primary consumers are microbial fungi and bacteria. These organisms, in turn, are preyed upon by meiofauna in the decaying grass, the surface microfilm of the marsh, and the decaying bases of plant shoots. Most of these microscopic organisms are protozoa, nematodes, harpacticoid copepods, annelids, rotifers, and larval stages of larger invertebrates. The larger invertebrates on the marsh surface are of two groups: foragers (deposit feeders) and filter feeders. In a general sense, they are considered aquatic because most have some kind of organ to filter oxygen out of water. Foragers include polychaetes, gastropod mollusks such as *Littorina irrorata* and *Melampus bidentatus*, and crustaceans such as *Uca* spp., the blue crab (*Callinectes sapidus*), and amphipods. These organisms browse on the sediment surface, ingesting algae, detritus, and meiofauna. The filter feeders such as the ribbed mussel (*Geukensia demissus*) and the oyster (*Crassostrea virginica*) filter particles out of the water column.

Aquatic Habitat

Animals classified as aquatic overlap with those in the benthic habitat. For convenience, we include in this group animals in higher trophic levels (mostly vertebrates) and migratory organisms that are not permanent residents of the marsh. Few fish species are permanent residents of the marsh. Most feed along the marsh edges and in small, shallow marsh ponds and move up into the marsh on high tides. Werme (1981) found 30% of silverside (*Menidia extensa*) and mummichog (*Fundulus heteroclitus*) in a North Atlantic estuary up in the marsh at high tide. Fish common in small salt marsh ponds in Louisiana include sheepshead minnow (*Cyprinodon variegatus*), diamond killifish (*Adinia xenica*), tidewater silverside (*Menidia beryllina*), gulf killifish (*Fundulus grandis*), and sailfin molly (*Poecilia latipinna*). Shrimp (*Penaeus* spp.) and blue crabs (*Callinectes sapidus*) are also common. Most other species use the marsh intermittently for shelter and for food but range widely. Many fish and shellfish spawn offshore or upstream and, as juveniles, migrate into the salt marsh, which offers an abundant food supply and shelter. As subadults, they migrate back into the estuary or offshore. This group of migratory organisms includes more than 90% of the commercially important fish and shellfish of the southeastern Atlantic and Gulf coasts.

Mammals

Two mammals in North American salt marshes deserve attention because of their impact on salt marshes: the muskrat and the coypu. The muskrat (*Ondatra zibethicus*) is native to North America; the coypu or nutria (*Myocastor coypus*) is an exotic species introduced from South America. Both prefer fresh marshes but are also found in salt marshes. In Louisiana, the muskrat appears to have been displaced by the nutria from its preferred freshwater habitat into saline marshes. Both mammals are voracious herbivores that consume plant leaves and shoots during the growing season and dig up tubers during the winter. They destroy far more vegetation than they ingest and are responsible for "eat-outs" that degrade large areas of marsh. These areas recover extremely slowly, especially in the subsiding environment of the northern Gulf Coast. In European marshes, it is common to have domestic animals (for example, cattle, sheep, or goats) grazing in coastal salt marshes (Bouchard et al., 2003). This grazing has a profound effect on the plant communities and zonation that develops in these marshes.

Ecosystem Function

Major points that have been demonstrated in several studies about the functioning of salt marsh ecosystems include the following:

- Primary productivity of macrophytes is high in much of the salt marsh—almost as high as in subsidized agriculture. This high productivity is a result of subsidies in the form of tides, nutrient import, and abundance of water that offset the stresses of salinity, widely fluctuating temperatures, and alternate flooding and drying.

- Although the biomass of edaphic algae is small, algal production can sometimes be as high as or higher than that of the community's macrophytes, especially in hypersaline marshes.

- Direct grazing of vascular plant tissue is a minor energy flow in the salt marsh, but grazing on edaphic and epiphytic algae is a significant source of high-quality food energy for meio- and macro-invertebrates.

- Fungi and bacteria are primary consumers that break down and transform indigestible plant cellulose (detritus) into protein-rich microbial biomass for consumers. This detrital pathway is a major flow of energy utilization in the salt marsh.

- Salt marshes have been shown at times to be both sources and sinks of nutrients, particularly nitrogen.

Primary Productivity

Tidal marshes are among the most productive ecosystems in the world, annually producing up to 80 metric tons per hectare of plant material ($8,000\,\mathrm{g\,m^{-2}\,yr^{-1}}$) in the southern Coastal Plain of North America. The three major autotrophic units of the salt marsh are marsh grasses, mud algae, and phytoplankton of the tidal creeks. Extensive studies of the net primary production have been conducted in salt marshes, especially along the Atlantic and Gulf coasts of the United States. A comparison of some of the measured values of net aboveground and belowground production is given in Table 2-3. Aboveground production varies widely, from as little as 410 $\mathrm{g\,m^{-2}\,yr^{-1}}$ in a Normandy salt marsh to a high of $4,200\,\mathrm{g\,m^{-2}\,yr^{-1}}$ in a Louisiana *Spartina patens* marsh. Belowground production is difficult to measure and can be much higher than aboveground production. Productivity of salt marshes is often higher along creek channels and in low or intertidal marshes than in high marshes because of the increased exposure to tidal and freshwater flow. These conditions also produce the taller forms of *Spartina*, as discussed earlier. Belowground production is sizable—often greater than aerial production (see Table 2-3). Under unfavorable soil conditions, plants seem to put more of their energy into root production. Hence, roots:shoot ratios seem to be generally higher inland than at streamside locations.

Sullivan and Currin (2000) summarized the productivity of edaphic algae. Annual benthic algal production, as measured in a number of studies, ranges from $28\,\mathrm{g\,C\,m^{-2}\,yr^{-1}}$ in a Gulf of Mexico coast *Juncus roemerianus* marsh to $341\,\mathrm{g\,C\,m^{-2}\,yr^{-1}}$

Table 2-3 Net Primary Productivity Estimates of Salt Marsh Plant Species in the Same Region

Species	Net Primary Production (g m^{-2} yr^{-1}) Aboveground	Belowground	Source
Louisiana			
Distichlis spicata	1,162–1,291		White et al. (1978)
Juncus roemerianus	1,806–1,959		
Spartina alterniflora	1,473–2,895		
Spartina patens	1,342–1,428		
Distichlis spicata	1,967		Hopkinson et al. (1980)
Juncus roemerianus	3,295		
Spartina alterniflora	1,381		
Spartina cynosuroides	1,134		
Spartina patens	4,159		
Alabama			
Juncus roemerianus	3,078	7,578	Stout (1978)
Spartina alterniflora	2,029	6,218	
Mississippi			
Juncus roemerianus	1,300		de la Cruz (1974)
Distichlis spicata	1,072		
Spartina alterniflora	1,089		
Spartina patens	1,242		
France (Normandy)			
Spartina anglica/ Salicornia/Suaeda maritima low marsh	1,080 (non-grazed) 410 (grazed)		Lefeuvre et al. (2000)
high marsh	1,990 (non-grazed) 550 (grazed)		

in a southern California *Jaumea carnosa* marsh (Table 2-4). Benthic algal production increases in a southerly direction along the Atlantic coast, but is lowest on the Gulf coast. Much of the algal production on the east and west coasts of the United States occurs when the overstory plants are dormant. On the Atlantic and Gulf coasts, the productivity of algae is 10 to 60% of vascular plant productivity. Zedler (1980), however, found that algal net primary productivity in southern California was 76 to 140% of vascular plant productivity. She hypothesized that the arid and hypersaline conditions of southern California favor algal growth over vascular plant growth. Algae are important components of the salt marsh food web, so much so that Kreeger and Newell (2000) stated: "We question the paradigm that salt marshes have 'detritus-based food webs' (Odum, 1980), considering that the bulk of

Table 2-4   Comparison of Annual Benthic Microalgal Production (g C m^{-2} yr^{-1}) and Ratio of Annual Benthic Microalgal to Vascular Plant Net Aerial Production (BMP/VPP) in Different Salt Marshes of the United States

State	Algal productivity, g C m^{-2} yr^{-1}	BMP/VPP, %	Reference
Massachusetts	105	25	Van Raalte (1976)
Delaware	61–99	33	Gallagher and Daiber (1974)
South Carolina	98–234	12–58	Pinckney and Zingmark (1993)
Georgia	200	25	Pomeroy (1959)
Georgia	150	25	Pomeroy et al. (1981)
Mississippi	28–151	10–61	Sullivan and Moncreiff (1988)
Texas	71	8–13	Hall and Fisher (1985)
California	185–341	76–140	Zedler (1980)

Source: Adapted from Sullivan and Currin, (2000).

secondary production by metazoans could actually be linked to primary production by the microphytobenthos rather than through either direct (herbivory) or indirect (detrivory) linkages to primary production by vascular plants."

Variations in productivity on the local scale result from complex interactions of soil anoxia, soluble sulfide, and salinity (Mendelssohn and Morris, 2000). Although water appears plentiful, the concentration of dissolved salt makes the salt marsh environment similar in many respects to a desert: The "normal" water gradient is from plant to substrate. To overcome the osmotic influence of salt, plants must expend energy to increase their internal osmotic concentration in order to take up water. As a result, numerous studies confirm that plant growth is progressively inhibited by increasing salt concentrations in the soil. This is true even for the salt-tolerant species of the salt marsh, and the salinity effects may be subtle. For example, Morris et al. (1990) showed that the year-to-year variation in marsh production at a single site on the East Coast was correlated with the mean summer water level, which they equated with soil salinity (soil salinity was inversely correlated with the frequency of marsh flooding in this study).

Another factor limiting production is the degree of anaerobiosis of the substrate. Vascular plants, even those that have developed adaptations to anaerobic conditions, grow best in aerobic soils. Many effects of anaerobiosis have been documented: reduced energy availability as the aerobic respiratory pathway is blocked, reduced nutrient uptake, the accumulation of toxic sulfides in the substrate, and changes in the availability of nutrients. Salt inhibition and oxygen depletion frequently occur together. *Spartina* grows shorter in the inland marsh because its drainage is poor; hence oxygen deficits are severe. Salt may concentrate in this environment. The primary result of poor drainage in inland salt marshes, however, is apparently a dramatically lower soil redox potential, which in turn leads to elevated sulfide concentrations. Although *S. alterniflora* is able to mitigate the toxic effects of sulfide to some extent through its ability to transport oxygen through the root system to the rhizosphere and by the enzymatic oxidation of sulfides, its growth is inhibited when the interstitial soluble sulfide concentration exceeds 1 mM sulfide (Bradley and Dunn, 1989; Koch et al., 1990).

Primary productivity also contributes to sediment accretion in salt marshes and there is increasing interest in the ability of salt marshes to withstand relative rises in sea level. Because of their location, salt marshes are constantly adjusting to maintain an equilibrium near mean sea level. Morris et al. (2002) demonstrated the importance of primary productivity for increased sediment accretion. By experimentally increasing productivity in a South Carolina salt marsh, sediment accretion in the marsh was also enhanced. Salt marshes tend to be most productive at elevations just below mean high tide (see Figure 2-10). However, in terms of long-term response to rapidly rising sea levels, these lower marshes may be incapable of accreting sediment quickly enough to keep pace with rising water levels. Marshes at slightly higher elevations with abundant supplies of sediment are the most likely to acclimate to rapidly rising sea levels.

Decomposition and Consumption

Since Teal's seminal publication on energy flow in the salt marsh system (Teal, 1962), salt marshes have been considered detritus systems. Almost three-quarters of the primary production in the salt marsh ecosystem is broken down by bacteria and fungi. In his study of energy flow within the salt marsh environment, Teal (1962) estimated that 47% of the total net primary productivity was lost through respiration by microbes. It was largely assumed that the rich secondary productivity of estuaries was fueled by a detritus food web. In recent years, with the development of new techniques such as multiple stable isotope fractionation, these early assumptions have been questioned,

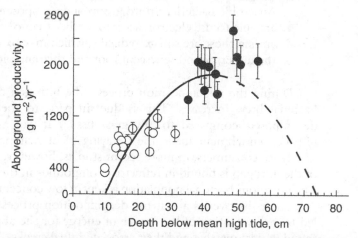

Figure 2-10 Aboveground net primary productivity (ANPP) as a function of mean sea level (as estimated by average annual water depth below mean high tide during peak growing season of June and July) for high (open circles) and low (solid circles) *Spartina alterniflora* salt marsh. (*After Morris et al., 2002*)

and a refined and quite different picture of decomposition and secondary production has emerged. With a few exceptions such as salt marshes in Mediterranean-type climates, primary production is dominated by emergent spermatophytes, usually grasses. When they senesce, the soluble organic contents are rapidly flushed from their tissues. This labile soluble organic matter from both living and decomposing salt marsh vegetation (which may be as much as 25% of the initial dry weight of the dying grass) is an important energy source for microorganisms in the marsh and the adjacent estuary (Wilson et al., 1986; Newell and Porter, 2000). The remaining 75% of dead vegetation biomass is largely composed of refractory structural lignocellulose that is indigestible by all but a few metazoans. Ideas about the fate of the vegetation biomass have changed. Some key conclusions about the process of decomposition are:

1. The initial secondary producers, or decomposers, on epibenthic marsh grass stems are ascomycetous fungi. These fungi may reach a biomass equal to 3 (summer) to 28 (winter) percent of live *Spartina alterniflora* standing crop. Most of this biomass occurs in standing dead grass or on the marsh surface. In South Atlantic coastal marshes, fungal productivity is 10 times greater in winter than in summer. In contrast, most of the bacterial biomass is found in the sediment surface microlayer. Productivity of bacteria is twice fungal productivity in summer, but only one-tenth as great in winter. The conversion efficiency of grass biomass to fungal biomass can be as high as 50% (Newell and Porter, 2000).

2. There appear to be at least three decomposer groups: (a) Fungi are the major decomposers of the epibenthic standing dead grass; (b) Aerobic bacteria in the surface microlayer decompose the decayed grass leaf shoots that are shredded by gastropods and amphipods and fall to the marsh surface; (c) Anaerobic bacteria, a third group of decomposers in deeper anoxic sediments, are able to use electron acceptors other than oxygen to metabolize. Prime among these are sulfate reducers, which may oxidize a major proportion of the underground senescent root and rhizome biomass.

During the decomposition process, the nitrogen content of the grass/fungal/ bacterial brew increases. This is due, in part, to the low C:N ratio of bacterial decomposers compared to raw grass tissue. It was assumed for many years that nitrogen enrichment made the decaying plant material a nutritionally better food supply for consumers; in more recent studies, however, it was determined that much of the nitrogen is bound in refractory compounds in the decaying grass (Teal, 1986). The nutritious bacterial population is kept at low concentrations by metazoan grazing.

These discoveries about the decomposition process in marsh macrophytes have led to a reevaluation of the source of energy for the abundant consumer population found in tidal marshes and their associated tidal creeks. Although much of the change consists of elaboration and clarification of the detrital process, a major shift has been toward a much greater role for algae, both phytoplankton and especially edaphic algae, as major flows of energy in the salt marsh food web. Vascular plants are still the major source of organic carbon, but few metazoans can assimilate this cellulose-rich material. Hence, direct grazers are limited to several species of herbaceous insects, which collectively consume less than 10% of plant production. The ribbed mussel *Geukensia demissa* is an exception to this generalization. It has been shown to assimilate aseptic detrital cellulose with an efficiency of up to 15%. Kreeger and Newell (2000) suggest that the mussel must either possess endogenous cellulases or contain a vigorous gut flora capable of cellulose breakdown.

Decay begins with fungal decomposition of the aerial parts of the senescent vascular plants. Epiphytic algae growing on the lower parts of the grass culms are also a part of this detrital brew. The complex is ingested and shredded by gastropod snails, such as *Littorina irrorata* and perhaps amphipods. In a microcosm experiment, the snails had the capacity to ingest 7% of their weight of naturally decayed leaves per day and assimilate them with an efficiency of about 50%. The epiphytic algae are also ingested and assimilated by amphipods and other organisms grazing on the dead leaf surfaces.

The finely shredded grass/fungal/algal material that falls to the marsh surface is infected by aerobic bacteria that continue the process of decomposition. Also part of this mixture is the algal community, largely diatoms, growing on the marsh surface. This complex is consumed by benthic meiofaunal and macrofaunal deposit feeders. Primary among the meiofauna are nematodes; also feeding on the surface are harpacticoid copepods, amphipods, polychaetes, turbellarians, ostracods, foraminifera, and gastroliths. The larger consumers in this group include fiddler crabs, snails, polychaetes, oligochaetes, and some bivalves.

Finally, some of the finely decomposed organic material on the surface microlayer is periodically suspended by winds and currents, where it mixes with the phytoplankton growing in the water. For example, as much as 25% of the suspended algae have been found to be edaphic species (MacIntyre and Cullen, 1995). This sestonic mix of bacterial/organic fragments, free-living bacteria, and algae is consumed by suspension feeders, especially benthic suspension feeders such as bivalve mollusks and oligochaete annelids. Also active are zooplankton, although they probably do not process as much material as the benthic bivalves. The meio- and macrofauna feeding on algae, fungi, and bacteria are in turn consumed by animals in the higher trophic levels (see Figure 2-9).

Organic Export

A central paradigm of salt marsh ecology has long been the *"outwelling hypothesis,"* which was first enunciated by E. P. Odum in 1968, based, in part, on a salt marsh energy flow analysis presented by John Teal at the first salt marsh conference, held in 1958 at the University of Georgia Marine Laboratory on Sapelo Island, Georgia (published as Teal, 1962). Odum described salt marshes as "primary production pumps" that feed large areas of adjacent waters (Odum, 1968), and he compared the flow of organic material and nutrients from salt marshes to the *upwelling* of deep ocean water, which supplies nutrients to some coastal waters. Teal (1962) hypothesized that salt marshes exported organic material and energy primarily as detritus from the marsh surface. In the intervening years, there have been many attempts to measure this export. Teal's (1962) energy flow analysis estimated that about 45% of net primary production was exported from the salt marsh. Nixon's (1980) summary agreed in that most studies showed an export of dissolved and particulate material, in an amount that could account for about 10 to 50% of phytoplankton production in coastal and estuarine waters.

Childers et al. (2000) pointed out that the original hypothesis was ambiguous in that it equated salt marsh export to coastal ocean import. In reality, the flows from a salt marsh are into nearby tidal creeks, and fluxes to the coastal ocean depend on the geomorphology of the estuary and the distance from the marsh to the coast. Hence, salt marshes interact with nearby tidal creeks and the inner estuary, which, in turn, exchange flows with the greater estuarine basin, which, finally, interacts with the coastal ocean (see Figure 2-11). Failure to take these spatial factors into account has made it difficult to compare studies and is one reason for the lack of agreement in study results.

The evidence for outwelling rests on more than organic flux data, such as reported in the studies summarized by Nixon (1980) and Childers (1994). Hopkinson (1985) reported that water column respiration offshore of the Georgia barrier islands exceeded *in situ* production; that is, the zone was heterotrophic, implying that organic matter was being imported from the inshore estuaries and marshes. Turner et al. (1979) reported that offshore, within 10 km of the coastal estuaries, primary productivity measurements were often 10 times greater than that farther offshore. They attributed this high productivity to outwelling of nutrients from the estuaries.

Other evidence of outwelling comes from fishery studies. Turner (1977) found a close correlation worldwide between commercial yield of shrimp (which are harvested

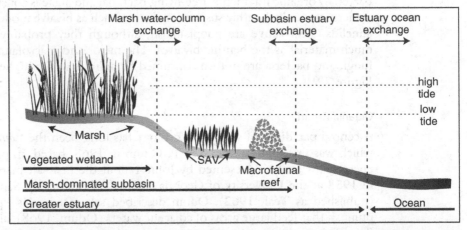

Figure 2-11 Hierarchy of estuarine-coastal landscape that includes estuarine subbasins nested within the greater estuary, and vegetated wetland ecosystems nested within both. (*After Childers et al., 2000*)

both in the estuary and offshore) and the area of estuarine intertidal vegetation. Teal and Howes (2000) analyzed fish catch statistics dating back to 1880 from the Long Island Sound, New York, and determined that fish catch was closely related to marsh edge length. Because edge length is an index of accessibility to the marsh, the result implicated salt marsh production in commercial fishery catch.

Several general factors affect the outwelling hypothesis. First, material and energy usually flow from concentrated hot spots to lower concentration areas. Salt marshes are hot spots of production, so it is logical to expect an outwelling of production and food energy (E. P. Odum, 2000). Second, outwelling can be expected to be modified by the geomorphology of the estuary and the location of a salt marsh in the estuary. Thus, open estuaries with salt marshes close to the coast are expected to export more material than estuaries with small coastal passes and distant marshes. Finally, salt marshes and coastal estuaries are pulsing systems, with daily tidal variation, seasonal variations in rainfall and river flow, and periodic severe storms. Extreme events often lead to import or export that overwhelms the normal daily fluxes.

Salt Marsh Die-Off

For the first several years of the new century (2000–2005), *Spartina* salt marshes in the southeastern and Gulf coasts of the United States were experiencing major die-off, totaling more than 100,000 ha and affecting 1500 km of coastline. One theory presented to explain this die-off was described by Silliman et al. (2005) as having the following sequence: a protracted and intense drought that occurred for 3 to 4 years (bottom-up effect) followed by snails (*Littoraria irrorata*) concentrating on the die-off borders to prolong the effect (top-down effect). In addition, declines in blue crab populations, a major predator of the snails, of 40 to 85% provided synergy for the snail grazing. Essentially it was suggested that "drought induced soil stress can amplify top-down control by grazers and initiate marsh plant die-off ... These disturbances then stimulate the formation of consumer fronts, leading to waves of salt marsh destruction resulting from runaway consumption" (Silliman et al., 2005). Such epidemic ecosystem die-offs that combine bottom-up and top-down stresses on coastal ecosystems in a synergistic way are another example of an undesirable positive feedback that could occur on coastal ecosystems with any significant climate change.

Tidal Freshwater Wetlands

Tidal freshwater wetlands are interesting because they receive the same "tidal subsidy" as mangroves and salt marshes but without the salt stress. One would expect, therefore, that these ecosystems might be very productive and also more diverse than their saltwater counterparts. As tides attenuate upstream, the wetlands assume more of the characteristics of inland freshwater wetlands (see Chapter 3). The distinction between freshwater tidal and inland wetlands is not clear-cut because on the coast they form a continuum (see Figure 2-2). Inland from the tidal salt marshes but still close enough to the coast to experience tidal effects, tidal freshwater marshes are dominated by a variety of grasses and by annual and perennial broad-leaved aquatic plants. In the United States, they are found primarily along the Middle and South Atlantic coasts and along the coasts of Louisiana and Texas. Tidal freshwater swamps tend to be most abundant along the furthest tidal extent of coastal rivers, particularly those rivers with low gradients and high discharge. Most of the extensive tidal freshwater forests in the United States occur along the southeastern coastline (Maryland to Texas). Estimates of tidal freshwater wetlands in the United States range from 400,000 ha along the Atlantic Coast to 819,000 ha for the conterminous United States (refer to Table 2-1). The extent of tidal freshwater swamps in the Unites States is less certain but 200,000 ha has been estimated for the southeast U.S. coastline (Field et al., 1991). The uncertainty in the estimates is related to where the line is drawn between tidal and non-tidal areas. Tidal freshwater marshes can be described as intermediate in the continuum from coastal salt marshes to freshwater marshes. Because they are tidally influenced but lack the salinity stress of salt marshes, tidal freshwater marshes have often been reported to be very productive ecosystems, although a considerable range in their productivity has been measured. Elevation differences across a freshwater tidal marsh correspond to different plant associations. These associations are not discrete enough to call communities, and the species involved change with latitude. Nevertheless, they are characteristic enough to allow some generalizations.

Vegetation

Vegetation is described here for both tidal freshwater marshes and tidal freshwater swamps. The species are generally common with nontidal freshwater marshes and swamps respectively. The hydroperiod, which fluctuates with twice-per-day tides in tidal freshwater wetlands rather than seasonally with nontidal freshwater wetlands, reflects an additional subsidy that generally enhances nutrient influx, plant growth, and organic export described in the Ecosystem Function section.

Marsh Vegetation

On the Atlantic Coast of the United States (see Figures 2-12a and b), submerged vascular plants such as *Nuphar advena* (spatterdock), *Elodea* spp. (waterweed), *Potamogeton* spp. (pondweed), and *Myriophyllum* spp. (water milfoil) grow in the streams and permanent ponds. The creek banks are scoured clean of vegetation each fall by the strong tidal currents, and they are dominated during the summer by annuals such as *Polygonum punctatum* (water smartweed), *Amaranthus cannabinus* (water hemp), and *Bidens laevis* (bur marigold). The natural stream levee is often dominated by *Ambrosia trifida* (giant ragweed). Behind this levee, the low marsh is populated with broad-leaved monocotyledons such as *Peltandra virginica* (arrow arum), *Pontederia cordata* (pickerelweed), and *Sagittaria* spp. (arrowhead).

Typically, the high marsh has a diverse population of annuals and perennials. W. E. Odum et al. (1984) called this the "mixed aquatic community type" in the Mid-Atlantic region. Leck and Graveline (1979) described a "mixed annual" association in New Jersey while Caldwell and Crow (1992) described the vegetation of

a.

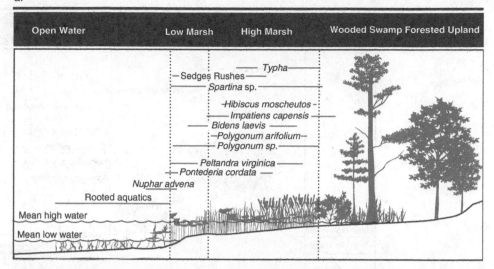

b.

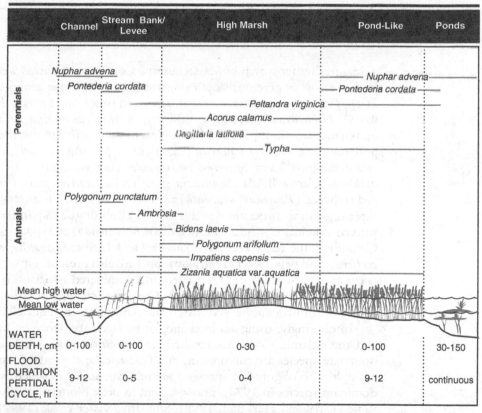

Figure 2-12   Cross sections across typical freshwater tidal marshes, showing elevation changes and typical vegetation: a. and b. Atlantic Coast marshes; c. new marsh in the Atchafalaya Delta, Louisiana. (*a. After W. E. Odum et al., 1984; b. After Simpson et al., 1983; c. After Gosselink et al., 1998*)

c.

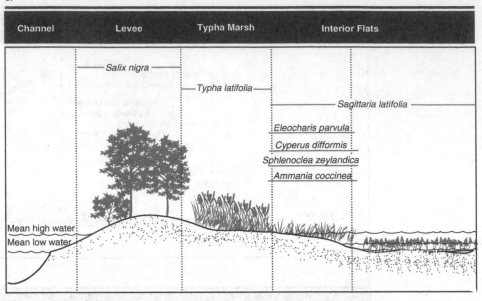

Figure 2-12   (continued)

a tidal freshwater marsh in Massachusetts. Generally, the areas were dominated early in the season by perennials such as arrow arum. A diverse group of annuals—*Bidens laevis, Polygonum arifolium* (tear-thumb) and other smartweeds, *Pilea pumila* (clear-weed), *Hibiscus coccineus* (rose mallow), *Acnida cannabina*, and others—assumed dominance later in the season. In addition to these associations, there are often almost pure stands of *Zizania aquatica* (wild rice), *Typha* spp. (cattail), *Zizaniopsis miliacea* (giant cutgrass), and *Spartina cynosuroides* (big cordgrass). In the northern Gulf of Mexico, arrowheads (*Sagittaria* spp.) replace arrow arum (*Peltandra* spp.) and pickerelweed (*Pontedaria cordata*) at lower elevations. Visser et al. (1998) described three vegetation associations in this area: (1) Bulltongue (*Sagittaria lancifolia*) occurs with co-dominants maidencane (*Panicum hemitomon*) and spikerush (*Eleocharis* spp.). Commonly the ferns *Thelypteris palustris* and *Osmunda regalis*, wax myrtle (*Myrica cerifera*), and pennywort (*Hydrocotyi* spp.) are also present. Fifty-two different species occur in this association. (2) A maidencane-dominated association is widespread across the delta, and includes 55 species. (3) Cutgrass (*Zizaniopsis miliacea*) occurring with co-dominant maidencane is relatively uncommon. It includes 20 other species.

Interestingly, rising sea level and/or surface subsidence on both the Gulf Coast and the Atlantic Coast has resulted in vegetation shifts. Although the previously dominant species are still present, in a Chesapeake Bay tidal freshwater wetland, for example, the oligohaline species *Spartina cynosuroides*, which was not among the dominant species in 1974, is now second in peak biomass and fourth in importance value (Perry and Hershner, 1999). Similarly, Visser et al. (1999) reported that the maidencane association has decreased from 51% coverage of the tidal wetlands of Terrebonne Bay (in the Mississippi River delta) in 1968 to only 14% in 1992. It has been replaced by *Eleocharis baldwinii*–dominated marshes, which were uncommon in 1968 (3% coverage) but in 1992 covered 42% of the area.

Floating Marshes Floating marshes in the tidal reaches of the northern Gulf of Mexico are similar to the nontidal riverine and lacustrine marshes found extensively around the globe. Large expanses of floating marshes (*Phragmites communis* marshes) have been found in the Danube Delta for at least a century (Pallis, 1915), along the lower reaches of the Sud in Africa (papyrus swamps; Beadle, 1974), in South America (floating meadows in lakes of the *varzea*; Junk, 1970), and in Tasmania (floating islands in the Lagoon of Islands; Tyler, 1976). Floating marshes have also been reported in Germany, the Netherlands (Verhoeven, 1986), England (Wheeler, 1980), and North Dakota and Arkansas in the United States (Eisenlohr, 1972; Huffman and Lonard, 1983). In Louisiana, floating marshes are usually floristically diverse, but different stands are dominated by *Panicum hemitomon* with ferns and vines such as *Vigna luteola* and *Ipomoea sagittata*; *Sagittaria lancifolia* with *Eleocharis* spp., *Panicum dichotomiflorum*, *Bacopa monnieri*, and *Spartina patens*; and *Eleocharis baldwinii* and *Eleocharis parvula* with *Ludwigia leptocarpa*, *Phyla nodiflora*, and *Bidens laevis* (Sasser et al., 1996). The marsh substrate is composed of a thick organic mat, entwined with living roots, that rises and falls (all year or seasonally) with the ambient water level (Swarzenski et al., 1991). This type of marsh is interesting in a successional sense because it appears to be an endpoint in development; it is freed from normal hydrologic fluctuation and mineral sediment deposition. Hence, in the absence of salinity intrusions, it appears to support a remarkably stable community (Sasser et al., 1995).

New Marshes The active deltas of the Mississippi and Atchafalaya Rivers are the sites of the largest newly prograding coastal deltas in the continental United States. Fresh tidal marshes (see Figure 2-12c) that have formed in the last 25 years on emergent islands are dominated on the natural levees by black willow (*Salix nigra*), and the extensive back-island mud flats by common three-square sedge (*Scirpus deltarum*) or by arrowhead (*Sagittaria latifolia*), with areas of cattail (*Typha latifolia*) and a seasonally variable annual/perennial mix in between.

Swamp Vegetation

Canopy trees richness tends to be low in tidal freshwater forests such as those found in southeastern United States. At the lowest elevations, canopy species are dominated by those that can withstand long periods of inundation such as cypress (*Taxodium distichum*), water tupelo (*Nyssa aquatica*), and swamp tupelo (*Nyssa biflora*) [see Chapter 3 for more descriptions of these tree species]. At slightly higher elevations, other species can become dominant, including ash (*Fraxinus* spp.), red maple (*Acer rubrum*), sweetgum (*Liquidambar styraciflua*), American hornbeam (*Carpinus caroliniana*), and sweetbay (*Magnolia virginiana*). Tidal freshwater swamps are often characterized by a distinctive hummock and hollow topography with most of the trees limited to the hummocks with hollows being sparsely vegetated by trees. Because these forests can have fairly open canopies, subcanopy and understory vegetation can be extensive and species-rich. In tidal swamps along the Pamunkey River in Virginia, Rheinhardt (1992) found spicebush (*Lindera benzoin*), winterberry (*Ilex verticillata*), *C. caroliniana*, and *Ilex opaca* to be among the most dominant subcanopy species. Understory dominants included halberd-leaved tearthumb (*Polygonum arifolium*), lizard's tail (*Saururus cernuus*), and sedges (*Carex* spp.). Along the forested tidal reaches of the Suwannee River on the Gulf Coast of Florida, dominant subcanopy vegetation was pumpkin ash (*Fraxinus profunda*), Carolina ash (*Fraxinus caroliniana*), and wax myrtle (*Morella cerifera*). Common understory vegetation included variable panic grass (*Dicanthelium commutatum*), string lily (*Crinum americanum*), *S. cernuus*, and *Carex* spp (Light et al., 2002).

Seed Banks

The species composition of a tidal freshwater wetland does not appear to depend on the availability of seed in particular locations. Seeds of most species are found in almost all habitats, although the most abundant seed reserves are generally from species found in that vegetation zone (Whigham and Simpson, 1975; Leck and Simpson, 1987, 1995; Baldwin et al., 1996). They differ, however, in their ability to germinate under the local field conditions and in seedling survival. Flooding is one of the main controlling physical factors. Many of the common plant species seem to germinate well even when submerged, for example, *Peltandra virginica* and *Typha latifolia*, whereas others, such as *Impatiens capensis*, *Cuscuta gronovii*, and *Polygonum arifolium*, show reduced germination. Baldwin et al. (2001) manipulated the hydrology of freshwater tidal marshes along the Patuxent River in Maryland and found that increasing flood depths by 3–10 cm can significantly reduce species richness and plant growth. In particular, shallow flooding early in the growing season reduced the germination of annuals.

Competitive factors also play a role in vegetation assemblages. Arrow arum and cattail, for example, produce chemicals that inhibit the germination of seed; and shading by existing plants is apparently responsible for the inability of arrow arum plants to become established anywhere except along the marsh fringes. Some species (*Impatiens capensis, Bidens laevis,* and *Polygonum arifolium*) are restricted to the high marsh because the seedlings are not tolerant of extended flooding. Seed bank strategies differ in different zones of the marsh. The seeds of most of the annuals in the high marsh germinate each spring so that there is little carryover in the soil. In contrast, perennials tend to maintain seed reserves. The seeds of most species, however, appear to remain in the soil for a restricted period. In one study, 31 to 56 percent of the seeds were present only in surface samples, and 29 to 52 percent germinated only in sediment samples taken in early spring (Leck and Simpson, 1987). The complex interaction of all these factors has not been elucidated to the extent that it is possible to predict what species will be established where on the marsh.

In addition to vascular plants, phytoplankton and epibenthic algae abound in freshwater tidal marshes, but relatively little is known about them. In one study of Potomac River marshes, diatoms (Bacillariophytes) were the most common phytoplankton, with green algae (Chlorophytes) comprising about one-third of the population and blue-green algae (Cyanobacteria) present in moderate numbers. The same three taxa accounted for most of the epibenthic algae. Indeed, many of the algae in the water column are probably entrained by tidal currents off the bottom. In a study of New Jersey tidal freshwater marsh soil algae, Whigham et al. (1980) identified 84 species exclusive of diatoms. Growth was better on soil that was relatively mineral and coarse, compared with growth on fine organic soils. Shading by emergent plants reduced algal populations in the summer months. In nontidal freshwater marshes, algae epiphytic on emergent plants and litter made important contributions to invertebrate consumers (Campeau et al., 1994). Algal biomass is probably two to three orders of magnitude less than peak biomass of the vascular plants, but the turnover rate is much more rapid.

Consumers

Coastal freshwater wetlands are used heavily by wildlife. The consumer food chain is predominantly detrital, and benthic invertebrates are an important link in the food web. Bacteria and protozoa decompose litter, gaining nourishment from the organic material. It appears unlikely that these microorganisms concentrate in large enough numbers to provide adequate food for macroinvertebrates. Meiobenthic organisms, primarily nematodes, comprise most of the living biomass of anaerobic sediments. They probably crop the bacteria as they grow, packaging them in bite-sized portions for slightly larger macrobenthic deposit feeders. In coastal freshwater marshes, the microbenthos is composed primarily of amoebae. This is in sharp contrast to more saline marshes in which foraminifera predominate. The slightly larger macrobenthos is composed of amphipods, especially *Gammarus fasciatus*, oligochaete worms, freshwater snails, and insect larvae. Copepods and cladocerans are abundant in the tidal creeks. The Asiatic clam (*Corbicula fluminaea*), a species introduced into the United States earlier in the century, has spread throughout the coastal marshes of the southern states. Caridean shrimp, particularly *Palaemonetes pugio*, are common, as are freshwater shrimp, *Macrobrachium* spp. The density and diversity of these benthic organisms are reported to be low compared with those in nontidal freshwater wetlands, perhaps because of the lack of diverse bottom types in the tidal reaches of the estuary.

Where coastal forests transition from tidal to non-tidal, shifts in invertebrate species have been shown to correspond with hydrological shifts. At the tidal transition of the Suwannee River, Wharton et al. (1982) reported that faunal associations changed from a brackish water snail-fiddler crab (*Neretina-Uca*) community to a freshwater snail-crayfish (*Vivipara-Cabarus*) community. Salinity and not vegetation was the primary factor dictating infauna and epifauna taxa along the lower reaches of the Cape Fear River in North Carolina (Hackney et al. 2007). Common faunal groups in tidal swamps include oligochaetes (especially Tubificidae and Lumbriculidae), fiddler crabs (*Uca* spp.) and grass shrimp (*Palaemontes pugio*).

Nekton

Coastal freshwater wetlands are important habitats for many nektonic species that use the area for spawning and year-round food and shelter and as a nursery zone and juvenile habitat. Fish of coastal freshwater marshes can be classified into five groups (see Figure 2-13). Most of them are freshwater species that spawn and complete their lives within freshwater areas. The three main families of these fish are cyprinds (minnows, shiners, carp), centrarchids (sunfish, crappies, bass), and ictalurids (catfish). Juveniles of all species are most abundant in the shallows, often using submerged marsh vegetation for protection from predators. Predator species, the bluegill (*Lepomis macrochirus*), largemouth bass (*Micropterus salmoides*), sunfish (*Lepomis* spp.), warmouth (*Lepomis gulosus*), and black crappie (*Pomoxis nigromaculatus*), are all important for sport fishing. Gar (*Lepisosteus* spp.), pickerel (*Esox* spp.), and bowfin (*Amia calva*) are other common predators often found in both coastal marshes and tidal freshwater creeks.

Some oligohaline or estuarine fish and shellfish that complete their entire life cycle in the estuary extend their range to include the freshwater marshes. Killifish (*Fundulus* spp.), particularly the banded killifish (*F. diaphanus*) and the mummichog (*F. heteroclitus*), are abundant in schools in shallow freshwater marshes, where they feed opportunistically on any available food. The bay anchovy (*Anchoa mitchilli*) and

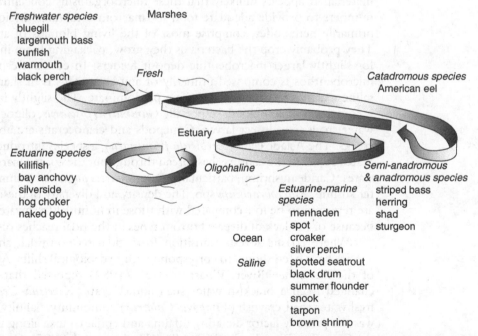

Figure 2-13   Fish and shellfish that use tidal freshwater marshes and other coastal systems can be classified into five groups: freshwater, estuarine, anadromous, catadromous, and estuarine-marine.

tidewater silverside (*Menidia beryllina*) are also often abundant in freshwater areas. The latter breed in this habitat more than in saltwater areas. Juvenile hog chokers (*Trinectes maculatus*) and naked gobies (*Gobiosoma bosci*) use tidal freshwater areas as nursery grounds (W. E. Odum et al., 1984).

Anadromous species of fish, which live as adults in the ocean, or *semi-anadromous species*, whose adults remain in the lower estuaries, pass through coastal freshwater marshes on their spawning runs to freshwater streams. For many of these species, the tidal freshwater areas are major nursery grounds for juveniles. Along the Atlantic Coast, herrings (*Alosa* spp.) and shads (*Dorosoma* spp.) fit into this category. The young of all of these species, except the hickory shad (*A. mediocris*), are found in peak abundance in tidal fresh waters, where they feed on small invertebrates and, in turn, are an important forage fish for striped bass (*Morone saxatilis*), white perch (*Morone americana*), catfish (*Ictalurus* spp.), and others (W. E. Odum et al., 1984). As they mature late in the year, they migrate downstream to saline waters and offshore. Two species of sturgeon (*Acipenser brevirostrum* and *A. oxyrhynchus*) were formerly important commercially in East Coast estuaries, but were seriously overfished and presently are rare. Both species spawn in nontidal and tidal fresh waters, and juveniles may spend several years there before migrating to the ocean.

The striped bass is perhaps the most familiar semi-anadromous fish of the Mid-Atlantic Coast because of its importance in both commercial and sport fisheries. Approximately 90% of the striped bass on the East Coast are spawned in tributaries of the Chesapeake Bay system. They spawn in spring in tidal fresh and oligohaline waters; juveniles remain in this habitat along marsh edges, moving gradually downstream to the lower estuary and nearshore zone as they mature. Because the critical period for survival of the young is the larval stage, conditions in the tidal fresh marsh area where these larvae congregate are important determinants of the strength of the year class.

The only *catadromous species* in Atlantic Coast estuaries is the American eel (*Anguilla rostrata*). It spends most of its life in fresh or brackish water, returning to the ocean to spawn in the region of the Sargasso Sea. Eels are common in tidal and nontidal coastal freshwater areas, in marsh creeks, and even in marshes.

The juveniles of a few species of fish that are marine spawners move into freshwater marshes, but most remain in the oligohaline reaches of the estuary. Species whose range extends into tidal freshwater marshes are menhaden (*Brevoortia tyrannus*), spot (*Leiostomus xanthurus*), croaker (*Micropogonias undulatus*), silver perch (*Bairdiella chrysoura*), spotted seatrout (*Cynoscion nebulosus*), black drum (*Pogonias cromis*), summer flounder (*Paralichthys dentatus*), snook (*Centropomus undecimalis*), and tarpon (*Megalops atlanticus*). Along the northern Gulf Coast, juvenile brown and white shrimp (*Penaeus* spp.) and male blue crabs (*Callinectes sapidus*) may also move into freshwater areas. These juveniles emigrate to deeper, more saline waters as temperatures drop in the fall.

Birds

Of all wetland habitats, coastal freshwater marshes may support the largest and most diverse populations of birds. W. E. Odum et al. (1984), working from a number of studies, compiled a list of 280 species of birds that have been reported from tidal freshwater marshes. They stated that although it is probably true that this environment supports the greatest bird diversity of all marshes, the lack of comparative quantitative data makes it difficult to test this hypothesis. Bird species include: waterfowl (44 species); wading birds (15 species); rails and shorebirds (35 species); birds of prey (23 species); gulls, terns, kingfishers, and crows (20 species); arboreal birds (90 species); and ground and shrub birds (53 species). A major reason for the intense use of these marshes is the structural diversity of the vegetation provided by broad-leaved plants, tall grasses, shrubs, and interspersed ponds.

Dabbling ducks (family Anatidae) and Canada Geese actively select tidal freshwater areas on their migratory flights from the north. They use the Atlantic Coast marshes in the late fall and early spring, flying farther south during the cold winter months. Most of these species winter in fresh coastal marshes of the northern Gulf, but some fly to South America. Their distribution in apparently similar marshes is variable; some marshes support dense populations, others few birds. For example, Fuller et al. (1988) found extensive use of new Atchafalaya River delta marshes by many species of ducks. Although the vegetation was dominated by arrowhead throughout the newly created islands, duck populations were twice as dense in the western islands of the delta compared to the east and central islands. On the central islands, ducks preferentially selected stands of three-square sedge over arrowhead; on the western islands where there was not three-square sedge, they frequented stands with mixed grass species over arrowhead. The reason for this selectivity is unclear. In the Atchafalaya Delta, it may be because the western islands remain fresh year-round, whereas the other islands sometimes experience saltwater encroachment (Holm, 1998). The birds feed in freshwater marshes on the abundant seeds of annual grasses and sedges and the rhizomes of perennial marsh plants and also in adjacent agricultural fields. They are opportunistic feeders, on the whole, ingesting from the available plant species. An analysis by Abernethy (1986) suggests that many species that frequent the fresh marsh early in the winter move seaward to salt marshes before beginning their northward migration in the spring. The reason for this behavior pattern is not known, but Abernethy speculated that the preferred foods of the freshwater marshes are depleted by early spring and the birds move into salt marshes that have not been previously grazed.

The Wood Duck (*Aix sponsa*) is the only duck species that nests regularly in coastal freshwater tidal marshes, although an occasional Black Duck (*Anas rubripes*) or Mallard (*A. platyrhynchos*) nest is found in Atlantic Coast marshes.

Wading birds are common residents of coastal freshwater marshes. They are present year-round in Gulf Coast marshes but only during the summer along the Atlantic Coast. An exception is the Great Blue Heron (*Ardea herodias*), which is seen throughout the winter in the northern Atlantic states. Nesting colonies are common throughout the southern marshes, and some species (Green-backed Herons (*Butorides striatus*) and Bitterns (*Ixobrychus exilis* and *Botaurus lentiginosus*) nest along the Mid-Atlantic Coast. They feed on fish and benthic invertebrates, often flying long distances each day from their nesting areas to fish.

Rails (*Rallus* spp.) and shorebirds, including the Killdeer (*Charadrius vociferus*), sandpipers (Scolopacidae), and the American Woodcock (*Scolopax minor*), are common in coastal freshwater marshes. They feed on benthic macroinvertebrates and diverse seeds. Gulls (*Larus* spp.), terns (*Sterna* spp.), Belted Kingfisher (*Ceryle alcyon*), and crows (*Corvus* spp.) are also common. Some are migratory; some are not. A number of birds of prey are seen hovering over freshwater marshes, including the Northern Harrier (*Circus cyaneus*), the American Kestrel (*Falco sparverius*), falcons (*Falco* spp.), eagles, Ospreys (*Pandion haliaetus*), owls (Tytonidae), vultures (Cathartidae), and the Loggerhead Shrike (*Lanius ludovicianus*). Swallow-tailed kites have been found to use the lower tidal reach of forested wetlands for nesting (Sykes et al., 1999). Arboreal birds use the coastal freshwater marshes intensively during short periods of time on their annual migrations. Flocks of tens of thousands of swallows (Hirundinidae) have been reported over the upper Chesapeake freshwater marshes. Flycatchers (Tyrannidae) are also numerous. They often perch on trees bordering the marsh, darting out into the marsh from time to time to capture insects. Although coastal marshes may be used for only short periods of time by a migrating species, they may be important temporary habitats. For example, the northern Gulf coastal marshes are the first landfall for birds on their spring migration from South America. Often, they reach this coast in an exhausted state and the availability of forested barrier islands for refuge and marshes for feeding is critical to their survival.

Sparrows, finches (Fringillidae), juncos (*Junco* spp.), blackbirds (Icteridae), wrens (Troglodytidae), and other ground and shrub birds are abundant residents of coastal freshwater marshes. W. E. Odum et al. (1984) indicated that 10 species breed in Mid-Atlantic Coast marshes, including the Ring-necked Pheasant (*Phasianus colchicus*), Red-winged Blackbird (*Agelaius phoeniceus*), American Goldfinch (*Carduelis tristis*), Rufous-sided Towhee (*Pipilo erythrophthalmus*), and a number of sparrows. The most abundant are the Red-winged Blackbirds, Dickcissels (*Spiza americana*), and Bobolinks (*Dolichonyx oryzivorus*), which can move into and strip a wild rice marsh in a few days.

Amphibians and Reptiles

Although W. E. Odum et al. (1984) compiled a list of 102 species of amphibians and reptiles that frequent coastal freshwater marshes along the Atlantic Coast, many are poorly understood ecologically, especially with respect to their dependence on this type of habitat. None is specifically adapted for life in tidal freshwater marshes. Instead, they are able to tolerate the special conditions of this environment. River turtles, the most conspicuous members of this group, are abundant throughout the southeastern United States. Three species of water snakes (*Nerodia*) are common. *Agkistrodon piscivorus* (the cottonmouth) is found south of the James River. In the South, especially along the Gulf Coast, the American alligator's preferred habitat is the tidal freshwater marsh. These large reptiles used to be listed as threatened or endangered, but they have come back so strongly in most areas that they are presently harvested legally (under strict control) in Louisiana and Florida. They nest along the banks of coastal freshwater marshes, and the animal, identified by its high forehead and long snout, is a common sight gliding along the surface of marsh streams.

Mammals

The mammals most closely associated with coastal freshwater marshes are all able to get their total food requirements from the marsh, have fur coats that are more or less impervious to water, and are able to nest (or hibernate in northern areas) in the marsh. These include the river otter (*Lutra anadensis*), muskrat (*Ondatra zibethicus*), nutria (*Myocastor coypus*), mink (*Mustela vison*), raccoon (*Procyon lotor*), marsh rabbit (*Silvilagus palustris*), and marsh rice rat (*Oryzomys palustris*). In addition, the opossum (*Didelphis virginiana*) and white-tailed deer (*Odocoileus virginianus*) are locally abundant. The nutria was introduced from South America some years ago and has spread steadily in the Gulf Coast states and into Maryland, North Carolina, and Virginia. It is not likely to spread farther north because of its intolerance to cold, but the South Atlantic marshes would seem to provide an ideal habitat. The nutria is more vigorous than the muskrat and has displaced it from the freshwater marshes in many parts of the northern Gulf. As a result, muskrat density is highest in oligohaline marshes. The muskrat, for some reason, is not found in coastal Georgia and South Carolina, or in Florida, although it is abundant farther north along the Atlantic Coast. Muskrat, nutria, and beaver (*Castor canadensis*) can influence the development of a marsh. The first two species destroy large amounts of vegetation with their feeding habits (they prefer juicy rhizomes and uproot many plants when digging for them), their nest building, and their underground passages. Beavers have been observed in tidal freshwater marshes in Maryland and Virginia. Their influence on forested habitats is well known, but their impact on tidal freshwater marshes needs to be studied more closely.

Ecosystem Function

Primary Productivity

Many production estimates have been made for freshwater coastal marshes. Productivity is generally high, usually falling in the range of 1,000 to 3,000 g m^{-2} yr^{-1} (see Table 2-5). The large variability reported from different studies stems, in part, from a lack of standardization of measurement techniques, but real differences can be attributed to several factors:

- **Type of plant and its growth habit:** Fresh coastal marshes, in contrast to saline marshes, are floristically diverse, and productivity is determined, at least to some degree, by genetic factors that regulate the species' growth habits. Tall perennial grasses, for example, appear to be more productive than broad-leaved herbaceous species such as arrow arum and pickerelweed.
- **Tidal energy:** The stimulating effect of tides on production has been shown for salt marshes and appears to be true for tidal freshwater marshes as well.
- **Other factors:** Soil nutrients, grazing, parasites, and toxins are other factors that can enhance or limit production in tidal freshwater marshes.

Table 2-5   Peak Standing Crop and Annual Net Primary Production (NPP) Estimates for Tidal Freshwater Marsh Associations in Approximate Order from Highest to Lowest Productivity

Vegetation Type[b]	Peak Standing Crop (g m^{-2})	Annual NPP (g m^{-2} yr^{-1})
Extremely high productivity		
Spartina cynosuroides (big cordgrass)	2,311	—
Lythrum salicaria (spiked loosestrife)	1,616	2,100
Zizaniopsis miliacea (giant cutgrass)	1,039	2,048
Panicum hemitomon (maidencane)	1,160	2,000
Phragmites communis (common reed)	1,850	1,872
Moderate productivity		
Zizania aquatica (wild rice)	1,218	1,578
Amaranthus cannabinus (water hemp)	960	1,547
Typha sp. (cattail)	1,215	1,420
Bidens spp. (bur marigold)	1,017	1,340
Polygonum sp./*Leersia oryzoides* (smartweed/rice cutgrass)	1,207	—
Ambrosia tirifida (giant ragweed)	1,205	1,205
Acorus calamus (sweet flag)	857	1,071
Sagittaria latifolia (duck potato)	432	1,071
Low productivity		
Peltandra virginica/Pontederia cordata (arrow arum/pickerelweed)	671	888
Hibiscus coccineus (rose mallow)	1,141	869
Nuphar adventa (spatterdock)	827	780
Rosa palustris (swamp rose)	699	—
Scirpus deltarium	—	523
Eleocharis baldwinii	130	—

[a]Values are means of one to eight studies.
[b]Designation indicates the dominant species in the association.
Source: W. E. Odum et al. (1984), Sasser and Gosselink (1984), Visser (1989), White (1993), and Sasser et al. (1995).

The elevation gradient across a fresh coastal marsh and the resulting differences in vegetation and flooding patterns account for three broad zones of primary production. The low marsh bordering tidal creeks, dominated by broad-leaved perennials, is characterized by apparently low production rates. Biomass peaks early in the growing season. Turnover rates, however, are high, suggesting that annual production may be much higher than can be determined from peak biomass. Much of the production is stored in belowground biomass (root:shoot $\gg$ 1) in mature marshes; this biomass is mostly rhizomes rather than fibrous roots. Decomposition is rapid, the litter is swept from the marsh almost as fast as it forms, the soil is bare in winter, and erosion rates are high. The parts of the high marsh dominated by perennial grasses and other erect, tall species are characterized by the highest production rates of freshwater species, and root: shoot ratios are approximately one. Because tidal energy is not as strong and the plant material is not so easily decomposed, litter accumulates on the soil surface, and little erosion occurs. The high-marsh mixed-annual association typically reaches a large peak biomass late in the growing season. Most of the production is above ground (root:shoot < 1), and litter accumulation is common.

Few estimates of primary productivity have been conducted in tidal freshwater swamps; however, they are expected to benefit from the same tidal nutrient subsidy. Relative elevation can have an effect on tree composition, geomorphology, flooding duration, and tidal exchange—all of which may influence productivity. There is likely a broad range of productivity and contribution for tidal freshwater wetlands. Tree growth in forests near their downstream threshold can be stunted by higher salinities and inundation frequency. Upstream, however, forests may be highly productive benefiting from the nutrient subsidy provided by the tides and diminished saltwater intrusion. Ozalp et al. (2007) found that aboveground net primary productivity (ANPP) at a tidal forest along the lower Pee Dee River in South Carolina ranged between 477 and $1117\,g\,m^{-2}\,yr^{-1}$. Along the tidal reaches of the Pamunkey River, Fowler (1987) found that forest ANPP was $1{,}230\,g\,m^{-2}\,yr^{-1}$, with 40% of this production from herbaceous plants. Along Lake Maurepas in southeastern Louisiana, where flooding durations and depths have increased due to anthropogenic alterations, Effler et al. (2007) reported that annual tree production of tidal swamps was low, ranging between 220 and $700\,g\,m^{-2}\,yr^{-1}$.

Energy Flow

There are three major sources of organic carbon to tidal freshwater marshes. The largest source is probably the vascular marsh vegetation, but organic material brought from upstream (terrestrial carbon) may be significant, especially on large rivers and where domestic sewage waters are present. Phytoplankton productivity is a largely unknown quantity. Most of the organic energy flows through the detrital pool and is distributed to benthic fauna and deposit-feeding omnivorous nekton. These groups feed fish, mammals, and birds at higher trophic levels. The magnitude of the herbivore food chain, in comparison to the detritus one, is poorly understood. Insects are more abundant in fresh marshes than in salt marshes, but most do not appear to be herbivorous. Marsh mammals apparently can "eat out" significant areas of vegetation (Evers et al., 1998), but direct herbivory is probably small in comparison to the flow of organic energy from destroyed vegetation into the detrital pool. Nevertheless, these rodents may exert strong control on species composition and on primary production (Evers et al., 1998). Herbivores also act in synergy with other stresses, for example, saltwater intrusion, flooding, and fire (Grace and Ford, 1996; Taylor et al., 1994).

The phytoplankton–zooplankton-juvenile fish food chain in fresh marshes is of interest because of its importance to humans. Zooplankton are an important dietary component for a variety of larval, postlarval, and juvenile fish of commercial importance that are associated with tidal freshwater marshes (Van Engel and Joseph, 1968).

Birds are major seasonal or year-round consumers in all types of tidal freshwater marshes. In addition, they move materials out of the system, processing it into guano, which in some areas may be a significant source of nutrients. They also modify plant composition and production by "eat-outs" (T. J. Smith and Odum, 1981).

Organic Export

In mature tidal freshwater marshes, most organic production is decomposed to litter and peat within the marsh system and nutrients are extracted and recycled. Floating tidal freshwater marshes may have even more closed cycles. Because they float, no surface flows export or import organic material. This limits fluxes to subsurface dissolved materials. The largest loss of organic energy in these mature marshes is probably to deep peats in the case of anchored marshes or to an organic sludge layer under the water column in floating marshes (Sasser et al., 1991). The magnitude of this loss was measured as 145 to $150\,\text{g C m}^{-2}\,\text{yr}^{-1}$ in a Gulf Coast freshwater marsh (Hatton, 1981).

Other losses of organic carbon from marshes occur through flushing from the marsh surface, conversion to methane that escapes as a gas, and export as biomass in the bodies of consumers that feed on the marsh. In highly reduced freshwater sediments, where, in contrast to salt marshes, little sulfur is available as an electron acceptor, it is expected that methanogenesis from carbon dioxide and fermentation should be a dominant pathway of respiratory energy flow. However, there is evidence that macrophyte biomass may regulate methanogenesis. Neubauer et al. (2005) examined the anaerobic metabolism in soils from tidal (freshwater and salt) marshes along the Patuxent River, Maryland. In the freshwater tidal marsh, they found that anaerobic metabolism was dominated by iron (III) reduction early in the growing season (see Figure 2-14). Later, when plant biomass declined, methanogenesis became the dominant pathway for anaerobic metabolism. The authors attributed the enhanced Fe(III) reduction to oxygen loss from plant roots during their peak growing period that allow the replenishment of Fe(III) oxides back into the rhizosphere. Rates of anaerobic metabolism were lower in the salt marsh and were dominated almost equally by both iron reduction and sulfate reduction when plants were most productive and almost exclusively by sulfate reduction late in the growing season. The relationship between plant biomass and Fe reduction was less clear in the salt marsh (see Figure 2-14) and may have been confounded by shifts in seasonal flooding.

The annual loss of carbon as methane from Gulf Coast freshwater marshes has been estimated as $160\,\text{g CH}_4\,\text{m}^{-2}\,\text{yr}^{-1}$ (C. J. Smith et al., 1982). In comparison, Lipschultz (1981) estimated a loss of only $10.7\,\text{g CH}_4\,\text{m}^{-2}\,\text{yr}^{-1}$ from a *Hibiscus*-dominated freshwater marsh in the Chesapeake Bay. Crozier and Delaune (1996) reported that methane production is correlated with labile sediment organic carbon. Because, on an areal basis, there is no clear gradient of labile carbon from fresh to salt marshes, their data did not show a clear consistent increase in methane production in freshwater marshes over saline ones.

Carbon dynamics in tidal freshwater forests can be variable and likely change along the estuarine gradient. Tidal export of detritus matter is likely significant; however,

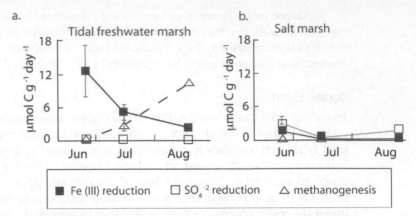

Figure 2-14 Seasonal changes in three anaerobic metabolism processes (iron reduction, sulfate reduction, and methanogeneis) in Maryland tidal freshwater and salt marshes. In the tidal freshwater marsh, iron reduction dominated early in the growing season (coinciding with peak plant biomass) and methanogenesis later. In the salt marsh, rates were lower and were dominated by iron and sulfate reduction early and by only sulfate reduction later in the growing season. *(After Neubauer et al., 2005)*

there is little information or estimates available for these wetlands. Soil organic matter tends to be high in these wetlands and is often linked to hydrology. A review of reported soil conditions in the southeastern U.S. found that the percentage of soil organic matter at the surface ranged from 9 to 77% with the highest concentrations reported for blackwater rivers (Anderson and Lockaby, 2007). For comparison, surface soils in a tidal freshwater shrub wetland in the Netherlands had 35% organic matter (Verhoeven et al., 2001). At the lower river reaches where waters can be more brackish, carbon mineralization and anaerobic metabolism may alternate between sulfate-reduction and methanogenesis, although iron reduction may be important as well. During low tides, water levels in these swamps can drop below the sediment surface and the increased aerobic conditions can increase methane oxidation and decrease emissions from these wetlands. During high tides, inflowing water can both cool soil temperatures and provide a medium for methane oxidation, reducing methane generation then as well (see Figure 2-15). Consequently, methane flux rates in tidal freshwater marshes and swamps tend to be lower than those in nontidal freshwater swamps (Anderson and Lockaby, 2007).

Nutrient Budgets

In general, nutrient cycling and nutrient budgets in coastal freshwater wetlands appear to be similar to salt marshes and mangrove swamps: they are fairly open systems that have the capacity to act as long-term sinks, sources, or transformers of nutrients. Even though these marshes generally are vigorously flooded by tides, they recycle a major portion of the nitrogen requirements of the vegetation. Figure 2-16 illustrates nitrogen cycling in a tidal freshwater marsh near Boston, Massachusetts. In this

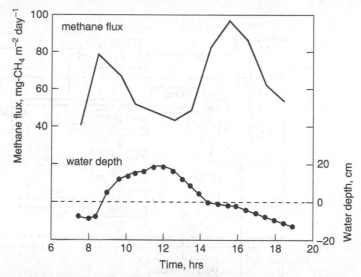

Figure 2-15 Methane fluxes from static chambers and concurrent tidal water level changes in a tidal freshwater swamp in White Oak River Estuary in North Carolina. The highest methane emission occurred while water levels coincided with the soil surface. When water levels were below the soil surface, methane oxidation increased in an aerobic surface layer resulting in decreased emission. When water levels exceeded the soil surface, methane emission was probably reduced by a water diffusion barrier. *(After Kelley et al., 1995)*

budget, most nutrient inputs are inorganic from the North River and the marsh and river nutrient cycles are mostly independent. Within the marsh itself, the major cycle is from peat to ammonium-nitrogen to live plants. Some of the ammonium-nitrogen is nitrified to nitrate-nitrogen and denitrified. Overall nitrate loss always exceeded denitrification measurements by acetylene block method, suggesting that other sinks of nitrate such as assimilatory nitrate reduction may be important (Bowden et al., 1991). Peat mineralization is sufficient to satisfy the nitrogen demands of the vegetation, and nearly all of the nitrogen flowing across the marsh from the adjacent river is re-exported. Mineralized litter and peat are conserved in the marsh by plant uptake and by microbial and litter immobilization. Despite this closed mineral cycle, Bowden et al. (1991) suggest that the small uptake of nitrogen from the river may be important.

Mangrove Wetlands

The coastal salt marsh of temperate middle and high latitudes gives way to its analog, the mangrove swamp, in tropical and subtropical regions of the world. The mangrove swamp is an association of halophytic trees, shrubs, and other plants growing in brackish to saline tidal waters of tropical and subtropical coastlines. This coastal, forested wetland (called a *mangal* by some researchers) is infamous for its

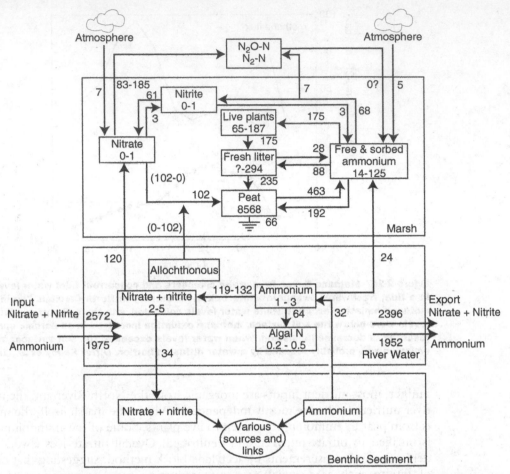

Figure 2-16   Nitrogen budget for a 22.8-ha tidal freshwater marsh in coastal Massachusetts, USA. Pool sizes are in kmoles N, fluxes in kmoles N yr^{-1}. (After Bowden et al., 1991)

impenetrable maze of woody vegetation, its unconsolidated peat that seems to have no bottom, and its many adaptations to the double stresses of flooding and salinity. The word *mangrove* comes from the Portuguese word *mangue* for "tree" and the English word *grove* for "a stand of trees" and refers to both the dominant trees and the entire plant community.

Many myths have surrounded the mangrove swamp. It was described at one time or another in history as a haven for wild animals, a producer of fatal "mangrove root gas," and a wasteland of little or no value. Researchers, however, have established the importance of mangrove swamps in exporting organic matter to adjacent coastal food chains, in providing physical stability to certain shore-lines to prevent erosion, in protecting inland areas from severe damage during hurricanes and tidal waves, and in serving as sinks for nutrients and carbon. There is extensive literature on the mangrove swamp on a worldwide basis—more than 5,000 titles. This interest probably stems from the worldwide scope of these ecosystems and the many unique features that they possess. Much of the early literature on mangroves concerned floristic and structural topics. Since the early 1970s, the focus has been on the functional aspects of mangrove swamps. Since that time, a significant literature on ecophysiology, primary productivity, stressors, food chains, and the detritus dynamics of mangrove ecosystems has been produced, along with some work on nutrient cycling, mangrove restoration, valuation of mangrove resources, and responses of mangroves to sea-level changes.

Geographical Extent

Mangrove swamps are found along tropical and subtropical coastlines throughout the world, usually between 25° N and 25° S latitude (see Figure 2-17a). Their limit in the Northern Hemisphere generally ranges from 24° to 32° N latitude, depending on the local climate and the southern limits of freezing weather. There are an estimated 240,000 km^2 of mangrove swamps in the world, with more than half of those swamps found in the latitudinal belts between 0° and 10° (see Figure 2-17b). Mangroves are divided into two groups—the Old World mangrove swamps and the New World and West African mangrove swamps. Over 50 species of mangroves exist, and their distribution is thought to be related to continental drift in the long term and possibly to transport by primitive humans in the short term. The distribution of these species, however, is uneven. The swamps are particularly dominant in the Indo-West Pacific region (part of the Old World group), where they contain the greatest diversity of species. There are 36 species of mangroves in that region, whereas there are only about 10 mangrove species in the Americas. It has been argued, therefore, that the Indo-Malaysian region was the original center of distribution for the mangrove species (Chapman, 1976b). Certainly some of the most intact mangrove forests in the world are found in Malaysia and in Micronesia, the small islands east of the Philippines in the western Pacific. Studies have illustrated how important these mangrove swamps are to local economies in these regions (Ewel et al., 1998; Cole et al., 1999). Several mangrove species, not native to the Hawaiian Archipelago despite its appropriate climate and coastal geomorphology, invaded the islands in the early 20th century and are now permanent fixtures on coastlines there (Allen, 1998).

There is also a great deal of segregation between the mangrove vegetation found in the Old World region and that found in the New World of the Americas and West Africa. Two of the primary genera of mangrove trees, *Rhizophora* (red mangrove) and *Avicennia* (black mangrove), contain separate species in the Old and New Worlds, suggesting "that speciation is taking place independently in each region" (Chapman, 1976b).

Florida Mangroves

Most of the mangrove swamps in the United States, estimated to cover 5,060 km^2, are found in Florida (refer to Table 2-1). The best development of mangroves in

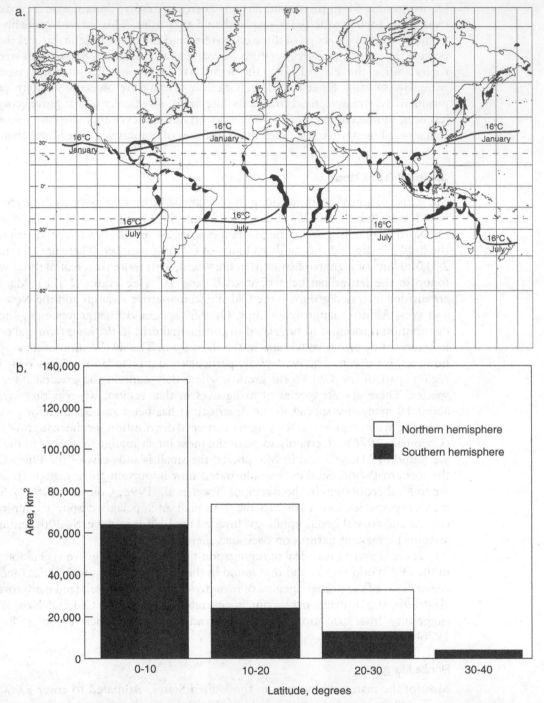

Figure 2-17 Distribution of mangrove wetlands: a. in the world, and b. by latitude. (*After Chapman, 1977; Twilley et al., 1992*)

Florida is along the southwest coast, where the Everglades and the Big Cypress Swamp drain to the sea. Mangroves extend up to 30 km inland along watercourses on this coast. The area includes Florida's Ten Thousand Islands, one of the largest mangrove swamps in the world at 600 km^2. Because of development pressure, a significant fraction of the original mangroves on these islands has been lost or altered. Patterson (1986) reported that there was a loss of 24% of mangroves on one of the most developed islands in this region, Marco Island, from 1952 to 1984. Mangroves are now protected in Florida, and it is illegal to remove them from the shoreline.

Mangrove swamps are also common further north along Florida's coasts, north of Cape Canaveral on the Atlantic and to Cedar Key on the Gulf of Mexico, where mixtures of mangrove and salt marsh vegetation appear. One species of mangrove (*Aviccnnia germinans*) is found in Louisiana and in the Laguna Madre of Texas, and it has been spreading extensively for the last 40 years, a possible sign of climate shift. Extensive mangrove swamps are also found throughout the Caribbean Islands, including Puerto Rico. Lugo (1988) estimated that there were originally 120 km^2 of mangroves in Puerto Rico, although only half of those remained by 1975.

Geographic Limitation by Frost and Competition

The frequency and severity of frosts are the main factors that limit the extension of mangroves beyond tropical and subtropical climes. For example, in the United States, mangrove wetlands are found primarily along the Atlantic and Gulf coasts of Florida up to 27° to 29° N latitude, north of which they are replaced by salt marshes. The red mangrove can survive temperatures as low as 2° to 4°C for 24 hours, whereas the black mangrove can withstand several days at this temperature, allowing black mangroves to extend farther north on Florida's east coast than red mangroves, as far north as 30° N. Similarly, Schaeffer-Novelli et al. (1990) described mangroves as extending to 28° to 30° S latitude along the Brazilian coast. Three to four nights of a light frost are sufficient to kill even the hardiest mangrove species. Lugo and Patterson-Zucca (1977) showed that mangroves survived approximately five nonconsecutive days of frost in January 1977 in Sea Horse Key Florida (latitude 29° N), but estimated that it would take 200 days for the forest to recover from frost damage. They also hypothesized that soil salinity stress could modify frost stress on mangroves, suggesting that the latitudinal limit of mangroves reflects a number of stresses rather than one factor.

Hydrogeomorphology

There are several different types of mangrove wetlands, each having a unique set of topographic and hydrodynamic conditions. A classification scheme of five geomorphological settings where mangrove forests occur, as developed by Thom (1982), includes systems dominated by waves, tides, and rivers, or most often by combinations of these three energy sources. As with the coastal salt marsh, the mangrove swamp can develop only where there is adequate protection from high-energy wave action. Several physiographic settings favor the protection of mangrove swamps, including

(1) protected shallow bays, (2) protected estuaries, (3) lagoons, (4) the leeward sides of peninsulas and islands, (5) protected seaways, (6) behind spits, and (7) behind offshore shell or shingle islands. Unvegetated coastal and barrier dunes usually develop where this protection does not exist, and mangroves are also often found behind these dunes.

In addition to the required physical protection from wave action, the range and duration of the flooding of tides exert a significant influence over the extent and functioning of the mangrove swamp. Tides constitute an important subsidy for the mangrove swamp, importing nutrients, aerating the soil water, and stabilizing soil salinity. Salt water is important to the mangroves in eliminating competition from freshwater species. Tides provide a subsidy for the movement and distribution of the seeds of several mangrove species. They also circulate organic sediments in some fringe mangroves for the benefit of filter-feeding organisms such as oysters, sponges, and barnacles and for deposit feeders such as snails and crabs. As with salt marshes, mangrove swamps are intertidal, although a large tidal range is not necessary. Most mangrove wetlands are found in tidal ranges of 0.5 to 3 m or more. Mangrove tree species can also tolerate a wide range of inundation frequencies. The red mangrove, *Rhizophora* spp., is often found growing in continually flooded coastal waters below normal low tide.

At the other extreme, mangroves can be found several kilometers inland along riverbanks where there is less tidal action. These mangroves depend on river discharge and are nourished by river flooding in addition to infrequent tidal inundation and the stability of groundwater and surface water levels near the coast.

Hydrodynamic Classification

The development of mangrove swamps is the result of topography, substrate, and freshwater hydrology, as well as tidal action. In the 1970s, Ariel Lugo, Sam Snedaker, and others at the University of Florida developed a classification of mangrove wetland ecosystems according to their physical and hydrologic conditions. The four major classes of mangrove wetlands, based on their hydrogeomorphology, are shown in Figure 2-18 and are discussed here.

Fringe mangroves—Fringe mangrove wetlands are found along protected shorelines and along some canals, rivers, and lagoons (see Figure 2-18a). They are common along shorelines that are adjacent to land higher than mean high tide but are exposed to daily tides. Fringe mangroves tend to accumulate organic debris because of the low-energy tides and the dense development of prop roots. Because the shoreline is open, these wetlands are often exposed to storms and strong winds that lead to the further accumulation of debris. Fringe mangroves are found on narrow berms along the coastline or in wide expanses along gently sloping beaches. If a berm is present, the mangroves may be isolated from freshwater runoff and would then have to depend completely on rainfall, the sea, and groundwater for their nutrient supply. A special case of fringe mangroves are small islands

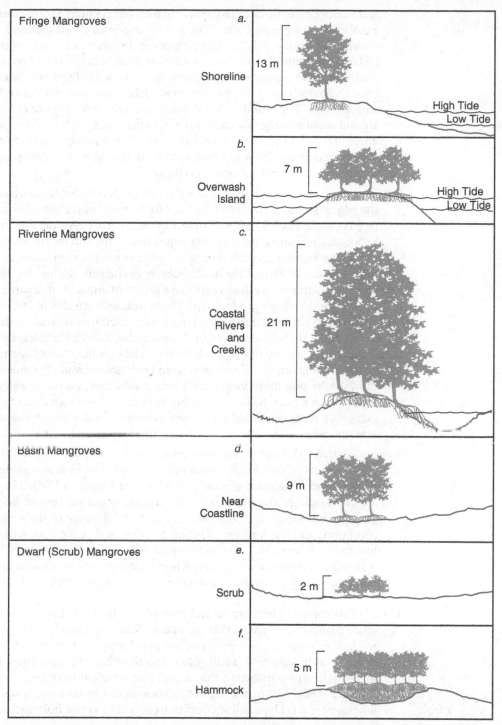

Figure 2-18 Classification of mangrove wetlands according to four hydrodynamic conditions and six types altogether: a. and b. fringe mangroves; c. riverine mangroves; d. basin mangroves; e. and f. scrub (dwarf) mangroves. (*After Wharton et al., 1976; Lugo, 1980; Cintrón et al. 1985*)

and narrow extensions of larger and masses (spits) that are "overwashed" on a daily basis during high tide. These are sometimes called *overwash mangrove islands* (see Figure 2-18b). The forests are dominated by the red mangrove (*Rhizophora*) and a prop root system that obstructs the tidal flow and dissipates wave energy during periods of heavy seas. Tidal velocities are high enough to wash away most of the loose debris and leaf litter into the adjacent bay. The islands often develop as concentric rings of tall mangroves around smaller mangroves and a permanent, usually hypersaline, pool of water. These wetlands are abundant in the Ten Thousand Islands region of Florida and along the southern coast of Puerto Rico. They are particularly sensitive to the effects of ocean pollution.

Riverine mangroves—Tall, productive riverine mangrove forests are found along the edges of coastal rivers and creeks, often several miles inland from the coast (see Figure 2-18c): These wetlands may be dry for a considerable time, although the water table is generally just below the surface. In Florida, freshwater input is greatest during the wet summer season, causing the highest water levels and the lowest salinity in the soils during that time. Riverine mangrove wetlands export a significant amount of organic matter because of their high productivity. These wetlands are affected by freshwater runoff from adjacent uplands and from water, sediments, and nutrients delivered by the adjacent river, and, hence, they can be significantly affected by upstream activity or stream alteration. The combination of adequate fresh water and high inputs of nutrients from both upland and estuarine sources causes these systems to be generally very productive, supporting large (16–26 m) mangrove trees. Salinity varies but is usually lower than that of the other mangrove types described here. The flushing of fresh water during wet seasons causes salts to be leached from the sediments (Cintrón et al., 1985).

Basin mangroves—Basin mangrove wetlands occur in inland depressions, or basins, often behind fringe mangrove wetlands, and in drainage depressions where water is stagnant or slowly flowing (see Figure 2-18d). These basins are often isolated from all but the highest tides and yet remain flooded for long periods when tidewater does flood them. Because of the stagnant conditions and less frequent flushing by tides, soils have high salinities and low redox potentials. These wetlands are often dominated by black mangroves (*Avicennia* spp.) and white mangroves (*Laguncularia* spp.), and the ground surface is often covered by pneumatophores from these trees.

Dwarf mangroves—There are several examples of isolated, low-productivity scrub mangrove wetlands that are usually limited in productivity because of the lack of nutrients or freshwater inflows. Dwarf mangrove wetlands are dominated by scattered, small (often less than 2 m tall) mangrove trees growing in an environment that is probably nutrient poor (see Figure 2-18e). The nutrient-poor environment can be a sandy soil or limestone marl. Hypersaline conditions and cold at the northern extremes of the mangrove's range can also produce "scrub," or stressed mangrove trees, in riverine, fringe, or basin wetlands. True dwarf mangrove wetlands, however, are found in the coastal fringe of the Everglades and the Florida Keys and along the northeastern coast of Puerto Rico. Some of these wetlands in the Everglades are inundated by seawater only during spring tides or storm surges and are often flooded by freshwater runoff in the rainy season. These types of wetlands are actually an intermix of small red mangrove trees with marsh vegetation such as sawgrass (*Cladium jamaicense*) and rush (*Juncus roemerianus*). Hammock mangrove wetlands also occur as isolated, slightly raised tree islands in the coastal fringe of the Florida Everglades and have characteristics of both basin and scrub mangroves. They are slightly raised as a result of the buildup of peat in what was once a slight depression in the landscape (see Figure 2-18f). The peat has accumulated from many years of mangrove productivity, actually raising the surface from 5 to 10 cm above the surrounding landscape. Because of this slightly raised level and the dominance by mangrove trees, these ecosystems look like the familiar tree islands or "hammocks" that are found throughout the Florida Everglades. They are different in that they are close enough to the coast to have saline soils and occasional tidal influences and, thus, can support only mangroves.

The preceding hydrogeomorphic classes of mangrove are broad categorizations and within these mangrove types there are likely to be subtypes that can be defined by specific hydrologic conditions. Knight et al. (2008) characterized three basin forest subtypes for *A. marina* forests using hydrology and forest structure data from the Coombabah Lake region in Southeast Queensland, Australia. These subtypes included a "deep basin" (characterized by ~50 cm standing water, ~3 tides yr^{-1}, and mature tree development), a "medium depth basin" (characterized by 15–30 cm standing water, 20–40 tides yr^{-1}, and intermediate tree development), and a "shallow basin" (characterized by 5–15 cm standing water, ~80 tides yr^{-1}, and recent tree establishment). Mangrove subtypes could be categorized in the other hydrogeomorphic classes as well.

Soils and Salinity

Soil salinity in mangrove ecosystems varies from season to season and with mangrove type (see Table 2-6). In riverine mangrove systems, the soil salinity is less than that of normal seawater because of the influx of fresh water. In basin mangroves, on the other hand, salinity can be well above that of seawater because of evaporative losses (>50 ppt). J. H. Davis (1940) summarized several major points about salinity in mangrove wetlands from his studies in Florida years ago that still hold today:

- There is a wide annual variation in salinity in mangrove wetlands.
- Salt water is not necessary for the survival of any mangrove species but only gives mangroves a competitive advantage over salt-intolerant species.

Table 2-6   Soil Salinity Ranges for Major Mangrove Types

Hydrodynamic Type	Soil salinity, ppt
Fringe mangroves	
Avicennia	59
Rhizophora	39
Riverine mangroves	10–20[a]
Basin mangroves	
Avicennia	>50
Laguncularia	low salinity
mixed forest	30–40

[a]Higher in dry season when less freshwater streamflow is available
Source: Adapted from Cintrón et al., 1985.

- Salinity is usually higher and fluctuates less in interstitial soil water than in the surface water of mangroves.
- Saline conditions in the soil extend farther inland than normal high tide because of the slight relief, which prevents rapid leaching.

A wide spatial and temporal range of salinity, from the constant seawater salinity of outer-coast overwash and mangrove swamps to the brackish water in coastal rivers and canals, is found in a region in Florida dominated by mangrove wetlands. Seasonal oscillations in salinity are a function of the height and duration of tides, the seasonality and intensity of rainfall, and the seasonality and amount of fresh water that enters the mangrove wetlands through rivers, creeks, and runoff. In Florida, summer wet-season convective storms and associated freshwater flow in streams and rivers, as well as an occasional hurricane in the late summer or early fall, lead to the dilution of salt water and the lowest salinity concentrations. Salinity is generally highest during the dry season, which occurs in the winter and early spring.

Soil Acidity

Mangrove soils are often acidic, although in the presence of carbonate as is often the case in south Florida, the soil pore water can be close to neutrality. The soils are often highly reduced, with redox potentials ranging from −100 to −400 mV. The highly reduced conditions and the subsequent accumulation of reduced sulfides in mangrove soils cause extremely acidic soils in many mangrove areas. Dent (1986, 1992) reported a measured accumulation of $10 \, kg \, S \, m^{-3}$ of sediment per 100 years in mangroves. When these soils are drained and aerated for conversion to agricultural land, the reduced sulfides, generally stored as pyrites, oxidize to sulfuric acid (Mitsch and Gosselink, 2007), producing what are known as "cat clays." These highly acidic soils make traditional agriculture difficult and are one of the reasons that, when mangrove swamps are converted to fish ponds, the ponds have a short lifetime before they are abandoned. Dent (1992) argued that the "dereclamation" of some previously "reclaimed" marginal coastal soils back to mangroves and salt marshes may be the best strategy for these acidic soils.

Vegetation

As is evident in coastal salt marshes, the stresses of waterlogged soils and salinity lead to a relatively simple flora in most mangrove wetlands, particularly when compared to their upland-neighboring ecosystem, the tropical rain forest. There are more than 50 species of mangroves throughout the world (Stewart and Popp, 1987), representing 12 genera in 8 families. Fewer than 10 species of mangroves are found in the New World, and only 3 species are dominant in the south Florida mangrove swamps—the red mangrove (*Rhizophora mangle* L.), the black mangrove (*Avicennia germinans* L., also named *A. nitida* Jacq.), and the white mangrove (*Laguncularia racemosa* L. Gaertn.). Buttonwood (*Conocarpus erecta* L.), although strictly not a mangrove, is occasionally found growing in association with mangroves or in the transition zone between the mangrove wetlands and the drier uplands. Different associations of mangrove plants dominate each of the hydrologic types of mangrove wetlands described previously. Red mangroves (*Rhizophora*), which contain abundant and dense prop roots, dominate fringe mangrove wetlands, particularly along the edges that face the open sea. Riverine mangrove wetlands are also numerically dominated by red mangroves, although they are straight trunked and have relatively few, short prop roots. Black (*Apicennia* spp.) and white (*Laguncularia* spp.) mangroves also frequently grow in these wetlands. Basin wetlands support all three species of mangroves, although black mangroves are the most common in basin swamps and hammock wetlands are mostly composed of red mangroves. Scrub mangrove wetlands are typically dominated by widely spaced, short (less than 2 m tall) red or black mangroves.

A comparison of the structural characteristics of the major hydrodynamic types of mangrove wetlands is provided in Table 2-7. These data were compiled from more than 100 mangrove research sites throughout the New World. Fringe mangroves generally have a greater density of large trees (> 10 cm dbh) compared to riverine and basin mangroves. Riverine wetlands, however, have the largest trees and, hence, a much greater basal area and tree height than do fringe or basin mangroves. The biomass of riverine mangroves is generally the highest, although data are difficult to compare because of the different methods and sample sizes used in various observations. Cintrón et al. (1985) reported a range of aboveground biomass for the Florida mangroves of 9 to 17 kg/m^2 for riverine mangroves and 0.8 to 15 kg/m^2 for fringe mangroves. Single measures of 0.8 kg/m^2 for a dwarf mangrove wetland and 9.8 kg/m^2 for a hammock mangrove (both in Florida) were also reported.

Zonation

In trying to understand the vegetation of mangrove wetlands, most early researchers were concerned with describing plant zonation and successional patterns. Some attempts were made to equate the plant zonation found in mangrove wetlands with

Table 2-7 Structural Characteristics of Canopy Vegetation for Major Mangrove Types[a]

Hydrodynamic Type	Number of Tree Species	Number of Trees (#/ha)		Basal Area (m²/ha)		Stand Height (m)	Aboveground Biomass (kg/m²)
		>2.5 cm dbh	>10 cm dbh	>2.5 cm dbh	>10 cm dbh		
Fringe mangroves	1.7 ± 0.1 (33)	4005 ± 642 (33)	852 ± 115 (31)	22.2 ± 1.5 (33)	14.6 ± 1.9 (31)	13.3 ± 2.6 (32)	0.8–15.9 (8)
Riverine mangroves	1.9 ± 0.1 (36)	1979 ± 209 (28)	661 ± 71 (32)	30.4 ± 3.5 (5)	32.6 ± 4.7 (32)	21.2 ± 4.8 (26)	1.6–28.7 (8)
Basin mangroves	2.3 ± 0.1 (31)	3599 ± 400 (31)	573 ± 102 (21)	18.5 ± 1.6 (31)	10.6 ± 2.2 (21)	9.0 ± 0.7 (31)	—

[a]Data are based on mangrove sites in Florida, Mexico, Puerto Rico, Brazil, Costa Rica, Panama, and Ecuador. Values are the average ± standard error (number of observations) except for aboveground biomass, which is the range (number of observations).
Source: Adapted from Cintrón et al. (1985).

successional seres. Lugo (1980), however, warned "zonation does not necessarily recapitulate succession because a zone may be a climax maintained by a steady or recurrent environmental condition." J. H. Davis (1940) is generally credited with the best early description of plant zonation in Florida mangrove swamps, especially in fringe and basin mangrove wetlands (see Figure 2-19). He hypothesized that the entire ecosystem was accumulating sediments and was, therefore, migrating seaward. Typically, *Rhizophora mangle* is found in the lowest zone, with seedlings and small trees sprouting even below the mean low tide in marl soils. Above the low-tide level but well within the intertidal zone, full-grown *Rhizophora* with well-developed prop roots predominate. There tree height is approximately 10 m. Behind these red mangrove zones and the natural levee that often forms in fringe mangrove wetlands, basin mangrove wetlands, dominated by black mangroves (*Avicennia*) with numerous pneumatophores, are found. Flooding occurs only during high tides. Buttonwood (*Conocarpus erecta*) often forms a transition between the mangrove zones and upland ecosystems. Flooding occurs only during spring tides or during storm surges, and soils are often brackish to saline.

The zonation of plants in mangrove wetlands led some researchers (for example, J. H. Davis, 1940) to speculate that each zone is a step in an autogenic successional process that leads to freshwater wetlands and, eventually, to tropical upland

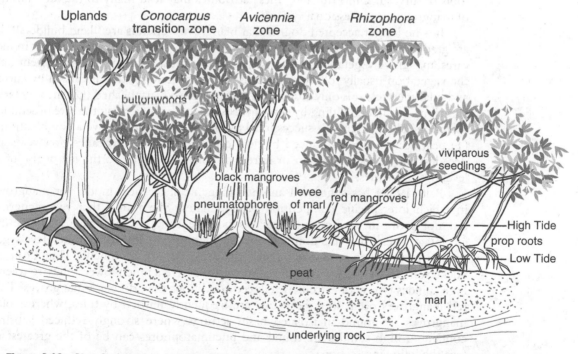

Figure 2-19 Classic zonation pattern of Florida mangrove swamp with illustrations of mangrove adaptations such as prop roots, viviparous seedlings, and pneumatophores.

forests or pine forests. Other researchers, led by Egler (1952), considered each zone to be controlled by its physical environment to the point that it is in a steady state or at least a state of arrested succession (allogenic succession). For example, with a rising sea level, the mangrove zones migrate inland, whereas during periods of decreasing sea level, the mangrove zones move seaward. Egler thought that the impact of fire and hurricanes made conventional succession impossible in the mangroves of Florida. Another theory, advanced by Chapman (1976b), is that mangrove succession may be a combination of both autogenic and allogenic strategies, or a "succession of successions." If that is the case, successional stages could be repeated a number of times before the next successional level is attained.

Lugo (1980) reviewed mangrove succession in light of E. P. Odum's criteria (1969; see Mitsch and Gosselink, 2007) and found that, except for mangroves on accreting coastlines, traditional successional criteria do not apply. Succession in mangroves is primarily cyclic, and it exhibits patterns of stressed or "youthful" ecosystems, including slowed or arrested succession, low diversity, $P:R$ greater than one, and open material cycles, even in mature stages. Lugo (1980) concluded that mangroves are true steady-state systems in the sense that they are the optimal and self-maintaining ecosystems in low-energy tropical saline environments. In such a situation, high rates of mortality, dispersal, germination, and growth are the necessary tools of survival. Unfortunately, these attributes may lead many to the identification of mangroves as successional systems.

It is no longer accepted dogma that mangrove wetlands are "land builders" that are gradually encroaching on the sea, as was suggested by J. H. Davis (1940). In most cases, mangrove vegetation plays a passive role in the accumulation of sediments, and the vegetation usually follows, not leads, the land building that is caused by current and tidal energies. It is only after the substrate has been established that the vegetation contributes to land building by slowing erosion and by increasing the rate of sediment accretion. The mangrove's successional dynamics appear to involve a combination of (1) peat accumulation balanced by tidal export, fire, and hurricanes over years and decades; and (2) advancement or retreat of zones according to the fall or rise of sea level over centuries.

Ball (1980) has suggested another allogenic succession model for mangroves, in which interspecific competition predominates. She found that red mangroves did not grow in dry upland locations because they did not have a competitive advantage there but did dominate where salinity and intertidal water levels gave them a competitive advantage. Using the same type of argument, Thibodeau and Nickerson (1986) suggested that red mangroves have a much lower ability to tolerate high sulfides typical of extremely reduced conditions than do black mangroves. Thus, red mangroves occur in regions that are frequently flushed by tides, whereas black mangroves are found in isolated basin settings where strongly reduced substrate containing high sulfides are found and pneumatophores can be of the greatest use (see the following section).

Mangrove Adaptations

Mangrove vegetation, particularly the dominant trees, has several adaptations that allow it to survive in an environment of high salinity, occasional harsh weather, and anoxic soil conditions. See Mitsch and Gosselink, 2007, Chapter 6, for an overview of wetland plant adaptations. These physiological and morphological adaptations have been of interest to researches and are among some of the most distinguishing features that the layperson notices when first viewing these wetlands. Some of the morphological adaptations are shown in Figure 2-20. Overall, physiological and morphological adaptations of mangroves include (1) salinity control, (2) prop roots and pneumatophores, and (3) viviparous seedling.

Salinity Control   Mangroves are facultative halophytes; that is, they do not require salt water for growth but are able to tolerate high salinity and, thus, out-compete vascular plants that do not have this salt tolerance. The ability of mangroves to live in saline soils depends on their ability to control the concentration of salt in their tissues. In this respect, mangroves are similar to other halophytes. Mangroves have

Figure 2-20   Adaptations of mangroves, including a. prop and drop roots of red mangroves in Senegal, West Africa, b. pneumatophores of black mangroves in southwest Florida, and c. viviparous propagule hanging in red mangrove canopy in southwest Florida. Photographs by W. J. Mitsch.

the ability both to prevent salt from entering the plant at the roots (*salt exclusion*) and to excrete salt from the leaves (*salt secretion*). Salt exclusion at the roots is thought to be a result of reverse osmosis, which causes the roots to absorb only fresh water from salt water. The root cell membranes of mangroves species of *Rhizophora*, *Avicennia*, and *Laguncularia*, among others, may act as ultrafilters that exclude salt ions. Water is drawn into the root through the filtering membrane by the negative pressure in the xylem developed through transpiration at the leaves; this action counteracts the osmotic pressure caused by the solutions in the external root medium. There are also a number of mangrove species (e.g., *Avicennia* and *Laguncularia*) that have salt-secreting glands on the leaves to rid the plant of excess salt. The solutions that are secreted often contain several percentages of NaCl, and salt crystals may form on the leaves. Another possible way, still questioned as to its importance, in which mangroves discharge salt is through leaf fall. This leaf fall may be significant because mangroves produce essentially two crops of leaves per year.

Prop Roots and Pneumatophores   Some of the most notable features of mangrove wetlands are the *prop roots* and *drop roots* of the red mangrove (*Rhizophora*; Figure 2-20a) and the numerous, small pneumatophores of the black mangrove (*Avicennia*) (reaching 20–30 cm above the sediments, although they can be up to 1 m tall; Figure 2-20b). The drop roots are special cases of the prop roots that extend from branches and other upper parts of the stem directly down to the ground, rooting only a few centimeters into the sediments. Oxygen enters the plant through small pores, called *lenticles*, which are found on both pneumatophores and prop and drop roots. When lenticels are exposed to the atmosphere during low tide, oxygen is absorbed from the air and some of it is transported to and diffuses out of the roots through a system of aerenchyma tissue. This maintains an aerobic microlayer around the root system. When prop roots or pneumatophores of mangroves are continuously flooded by stabilizing the water levels, those mangroves that have submerged pneumatophores or prop roots soon die.

In an interesting experiment to determine the importance of oxygen transport from the aerial organs to the sediments, Thibodeau and Nickerson (1986) "capped" with plastic tubing the pneumatophores of *Avicennia germinans* in a fringe mangrove forest in the Bahams. They observed that the aerobic zone surrounding the roots was reduced in the area by capping, indicating that the pneumatophores help the plant produce an oxidized rhizosphere. They also found that the greater the number of pneumatophores present in a given area, the more oxidized the soil. They described the relationship as

$$E_H = -307 + 1.1 \text{ pd} \tag{2.1}$$

where

E_H = redox potential (mV)

pd = pneumatophore density (number per 0.25 m^2)

Viviparous Seedlings   Red mangroves (and related genera in other parts of the world) have seeds that germinate while they are still in the parent tree; a long, cigar-shaped hypocotyl (*viviparous seedling*) develops while hanging from the tree (see Figure 2-20c). This is apparently an adaptation for seedling success where shallow anaerobic water and sediments would otherwise inhibit germination. The seedling (or propagule) eventually falls and often will root if it lands on sediments or will float and drift in currents and tides if it falls into the sea. After a time, if the floating seedling becomes stranded and the water is shallow enough, it will attach to the sediments and root. Often, the seedling becomes heavier with time, rights itself to a vertical position, and develops roots if the water is shallow. It is not well understood whether contact with the sediments stimulates root growth or if the soil contains some chemical compound that promotes root development. The value of the floating seedlings for mangrove dispersal and for invasion of newly exposed substrate is obvious. Rabinowitz (1978) reported that the obligate dispersal time (the time required during propagule dispersal for germination to be completed) was 40 days for the red mangrove and 14 days for the black mangrove propagules. She also estimated that red and black mangrove propagules could survive for 110 and 35 days, respectively.

Consumers

W. E. Odum et al. (1982) reported the following data from the literature describing faunal use of mangroves in Florida in terms of the number of species: 220 fish; 181 birds, including 18 wading birds, 29 water birds, 20 birds of prey, and 71 arboreal birds; 24 reptiles and amphibians; and 18 mammals. In general, a wide diversity of animals is found in mangrove wetlands; their distribution sometimes parallels the plant zonation described previously. Many of the animals that are found in mangrove wetlands are filter feeders or detritivores, and the wetlands are just as important as a shelter for most of the resident animals as they are a source of food. Some of the important filter feeders found in Florida mangroves include barnacles (*Balanus eburneus*), coon oysters (*Ostrea frons*), and the eastern oyster (*Crassostrea virginica*). These organisms often attach themselves to the stems and prop roots of the mangroves within the intertidal zone, filtering organic matter from the water during high tide.

Crabs are among the most important animal species in mangrove wetlands around the world, and they appear to play an important role in maintaining biodiversity in mangrove ecosystems (Twilley et al., 1996). They burrow in the sediments, prey on mangrove seedlings, facilitate litter decomposition, and are the key transfer organism for converting detrial energy to wading birds and fish in the mangrove forest itself and to offshore estuarine systems. Mangroves around the world are dominated by six of the 30 families of Brachyura that collectively make up about 127 species (Jones, 1984). *Uca* and *Sesarma* are the most abundant crab genera in mangrove wetlands. *Uca* spp. (fiddler crabs) are particularly abundant in mangrove wetlands in Florida, living on the prop roots and high ground during high water and burrowing in the sediments during low tide. *Sesarma* (sesarmids) is the most abundant genus of crabs in the world, with dozens of species in the Indo-Malaysian and East African mangroves, but many fewer in tropical America.

One of the most significant ways in which crabs may influence the distribution of mangroves is by selective predation of mangrove propagules (T. J. Smith, 1987; T. J. Smith et al., 1989; Obsborne and Smith, 1990). Yet, this effect is not common to all mangrove wetlands. Smith et al. (1989), in comparing the effect of this predation on mangroves around the world, found that crabs consumed up to 75% of the mangrove propagules in Australian mangrove swamps, but very little of the litterfall or propagules in Panamanian and Florida mangrove swamps.

The role of crabs in leaf litter removal (burial and consumption) has been illustrated in a number of studies in Malaysia (Malley, 1978), Australia (Robertson, 1986; Robertson and Daniel, 1989; T. J. Smith et al., 1991), and East Africa (Micheli et al., 1991). Robertson and Daniel (1989) estimated that in some mangrove forests, leaf processing by Sesarmid crabs alone was over 75 times the rate as microbial degradation and that crabs removed more that 70% of the litter of the high-intertidal *Ceriops* and *Bruguiera* mangroves in tropical Australia. Smith et al. (1991) developed an experiment where crabs were removed from experimental plots in *Rhizophora* forests. They found that the removal of crabs caused significantly higher concentrations of sulfide and ammonium in the mangrove soils, due primarily to the absence of burrowing, which oxygenates the soil. Taking into account the role that they have in seedling survival, carbon cycling, sediment microtopography, and soil chemistry, Smith et al. (1991) suggested that crabs are the keystone species of the mangrove ecosystem.

Many other invertebrates, including snails, sponges, flatworms, annelid worms, anemone, mussels, sea urchins, and tunicates, are found growing on roots and stems in and above the intertidal zone. Wading birds frequently found in Florida mangroves include the Wood Stork (*Mycteria americana*), White Ibis (*Eudocimus albus*), Roseate Spoonbill (*Ajaia ajaja*), cormorant (*Phalacrocorx* spp.), Brown Pelican (*Pelicanus occidentalis*), egrets, and herons (W. E. Odum and McIvor, 1990). Vertebrates that inhabit mangrove swamps include alligators, crocodiles, turtles, bears, wildcats, pumas, and rats.

Ecosystem Function

Certain functions of mangroves, such as net primary productivity, organic export, and outwelling have been studied extensively. A picture of the dynamics of mangrove wetlands has emerged from several key studies. These studies have demonstrated the importance of the physical conditions of tides, salinity, and nutrients to these wetlands and have shown where natural and human-induced stresses have caused the most effect.

Primary Productivity

Based on a global assessment of mangrove productivity, Bouillon et al. (2008) conservatively estimated worldwide mangrove productivity at 218 ± 72 Tg C yr^{-1} and added that over half of the C is unaccounted for, based on various estimates of mangrove C-sinks (organic export, burial, and mineralization). A wide range of

Table 2-8 Mass Balance of Carbon Flow (g C m^{-2} yr^{-1}) in Mangrove Forests in Florida and Puerto Rico

	Rookery Bay, Florida[a]		Puerto Rico Fringe[b]	Fahkahatchee Bay, Florida[c]		
	Fringe	Basin		Basin	Fringe	Fringe
Gross primary productivity (GPP)						
Canopy	2,055	3,292	3,004	3,760	4,307	5,074
Algae	402	26	276			
Total	2,457	3,318	3,280			
Respiration (plants)						
Leaves, stems	671	2,022	1,967	1,172	1,416	3,084
Roots, above ground	22	197	741	146	182	215
Roots, below ground	?	?	?	?	?	?
Total	693	2,219	2,708	1,318	1,598	3,299
Net primary production	1,764	1,099	572	2,442	2,709	1,775
Growth		186	153			
Litterfall		318	237			
Respiration (heterotrophs)		197	135			
Respiration (total)		2,416	2,843			
Export		64	500			
Net ecosystem production (NEP)		838	−63			
Burial		?	?			
Growth		186	153			

[a]Lugo et al. (1975), Twilley (1982, 1985), and Twilley et al. (1986).
[b]Golley et al. (1062).
[c]Carter et al. (1973).
Source: Adapted from Twilley (1988).

productivity has been measured in mangrove wetlands due to the wide variety of hydrodynamic and chemical conditions encountered. Table 2-8 presents a balance of carbon flow in several fringe and basin mangrove swamps in Florida and Puerto Rico. Net primary productivity ranges from approximately 570 to 2,700 g-C m^{-2} yr^{-1}. Gross and net primary productivity are highest in riverine mangrove wetlands, lower in fringe mangrove wetlands, and lowest in basin mangrove wetlands. The highest productivity in riverine mangrove wetlands is due to the greater influence of nutrient loading and freshwater turnover at the riverine site.

The important factors that control mangrove function in general and primary productivity in particular are: (1) tides and storm surges; (2) freshwater discharge; (3) parent substrate; and (4) water-soil chemistry, including salinity, nutrients, and turbidity. These factors are not mutually exclusive, for tides influence water chemistry and, hence, productivity by transporting oxygen to the root system, by removing the buildup of toxic materials and salt from the soil water, by controlling the rate of sediment accumulation or erosion, and by indirectly regenerating nutrients lost from the root zone. The principal chemical conditions that affect primary productivity are soil water salinity and the concentration of major nutrients. High soil salinities, which, in turn, are a function of the local hydrology and geomorphology, appear to be the most important variable that influences the productivity of mangroves in a given region. For example, one study of mangroves in Puerto Rico (Cintrón et al., 1978) found that tree height of the mangroves, as a measure of productivity, was inversely related to soil salinity according to the following relationship:

$$h = 16.6 - 0.20C_s \qquad (2.2)$$

where

h = tree height (m)
C_s = soil salinity concentration (ppt)

In a similar analysis, J. W. Day et al. (1996) were able to demonstrate a relationship between total litterfall in an *Avicennia*-dominated basin mangrove forest in Mexico and soil salinity as

$$L = 3.915 - 0.039C_s \qquad (2.3)$$

where

L = litterfall (g m^{-2}day^{-1})

In the Dominican Republic, Sherman et al. (2003) found strong correlations between several measures of productivity and soil salinity within a mature, even-aged mangrove forest. One exception was belowground biomass, which tended to increase with salinity, although the relationship was weak. This finding was supported by Lovelock (2008), who compared mangrove productivity from 11 forests around the world. Lovelock found very good correlations between soil respiration and measures of leaf production and biomass. However, the highest belowground carbon allocation per unit litterfall was found in scrub mangrove forests. It seems that mangroves allocate more growth belowground during environmentally stressful conditions.

Nutrient availability has also been shown to have a major influence on mangrove productivity. Geomorphic setting can often dictate nutrient availability and overall structure of mangrove swamps. However, with an increasing human presence in coastal areas, there is increased potential for nutrient enrichment from anthropogenic sources. Feller et al. (2007) experimentally relieved nutrient deficiencies in mangroves forests that were P-limited (in Twin Cays, Belize) and N-limited (Indian River Lagoon, Florida). When nutrient deficiency was relieved, black mangroves (*A. germinans*) at both forests responded with enhanced stem growth with the greatest response coming from the N-limited forest. Nutrient enrichment was found to influence more ecological processes in the P-limited Twin Cays forest and the authors concluded that eutrophication is more likely to shift nutrient limitations there than in the N-limited Indian River Lagoon.

Hurricane Effects

Hurricanes (and typhoons) and mangrove swamps could be described as having a turbulent tropical love-hate relationship. Either by genetic design or chance, the time required for the attainment of a steady state in mangrove forest in Florida is approximately the same as the average period between tropical hurricanes (approximately 20 to 24 years for Caribbean systems). This match suggests that mangroves may have adapted or evolved to go through one life cycle, on average, between major tropical storms. On average, a mangrove forest reaches maturity just as the next hurricane or typhoon hits.

When Hurricane Andrew passed over south Florida in the late summer of 1992, an opportunity existed for detailed studies of the immediate impact of hurricanes on mangroves. Major damage to mangroves in the vicinity of the Everglades occurred due to trunk snapping and uprooting rather than due to any storm surge (T. J. Smith et al., 1994). Mortality was greatest for red mangrove in the 15- to 30-cm DBH age class and over a wider band of 10 to 35 cm dbh for black and white mangrove. Two other interesting observations were made (T. J. Smith et al., 1994):

- Gaps that developed in the mangrove forests due to lightning prior to the hurricane were, after the hurricane, small green patches amid the gray matrix of dead mangroves. Apparently, the small-sized mangrove trees that were spared in these patches would now serve as propagule regeneration sites for the much larger scale area of catastrophic disturbance. Small-scale disturbance nested in large-scale disturbance provides a positive feedback for more rapid mangrove recovery.

- The loss of mangrove trees in a hurricane removes a major source of aeration of mangrove soils, the trees themselves. As a result of the loss of this aeration, soils in hurricane-impacted mangrove wetlands might become even more reduced, producing even more toxic hydrogen sulfide that could preclude mangrove regeneration for a number of years. Recovery of the mangrove forest to a system similar to that prior to the disturbance is not assured because of this negative feedback.

Other research after Hurricane Andrew has looked at the vegetative response. Monitoring forest plots inside and outside the eye-wall path of Hurricane Andrew between 1995 and 2005, Ward et al. (2006) found that turnover rates (mortality and recruitment) between the two sections differed (dynamic turnover rates were greater inside the eye-wall). However, both sections also exhibited a steady rate of forest turnover suggesting that ecological conditions and not structural conditions primarily control productivity.

In the case of a large climatic disturbance such as hurricanes, the deposition of coarse woody debris (CWD) may constitute a substantial nutrient flux. Using live stems of *A. germinans*, *L. racemosa*, and *R. mangle* in south Florida, Romero et al. (2005) found that decomposition of CWD was initially rapid during the decay of labile components ($0.37–23.71\%$ month^{-1}) and then slowed considerably as refractory components decayed (0.001–0.033% mo^{-1}). Highest decomposition rates were detected in *A. germinans* and were attributed to the higher proportion of labile tissue in that species. They found that CWD was initially a source of N but after two years became a sink. Using published estimates of CWD that would have been available per species following Hurricane Andrew (Smith et al., 1994, Krauss et al., 2005), the authors estimated the release of 1.34 Mg N ha^{-1} and 0.129 Mg P ha^{-1} in two months. However, between months 2-13, dead wood accumulated approximately 9 Mg N ha^{-1}.

Response to Rising Sea Levels

Because sea-levels are rising around the world, many coastal ecosystems may be adversely affected. It is unclear whether the rising sea levels may displace mangrove forests because these ecosystems naturally adapt and are resilient to disturbance. Alongi (2008) pointed out that mangroves have several characteristics that make them resilient to disturbance that are acute (e.g., hurricanes, tsunamis) or chronic (changing sea level). These characteristics include: belowground reservoirs of nutrients, rapid nutrient flux and decomposition, complex and efficient biotic controls, and often rapid reconstruction following disturbance due to the self-design and simple architecture of these forests.

It appears that some mangrove forests may be more capable of acclimating to a rising sea level than others. Alongi (2008) predicted that the most susceptible mangrove forests will be those slower growing forests at the thresholds of their habitat range, for instance, mangrove forests in arid regions where mangroves grow slower due to higher salinities, lower humidity, and extreme light conditions. Other susceptible mangroves will include forests in carbonate environments where growth is slow and the input of terrestrial sediment is limited. As a result, many of the world's mangrove forests are vulnerable to climate change including those in the Caribbean and the Pacific Islands. Based on current projections of climate change and sea level rise, Alongi (2008) predicted a decline of 10–15% of the global mangrove area by 2100. Although this decline is considerable, the author pointed out that the threat may be moot if current mangrove deforestation trends continue worldwide.

Export of Organic Material

Mangrove swamps are important exporters of organic material to the adjacent estuary through the same *outwelling* discussed for tidal salt marshes. In one of the first studies on outwelling from mangroves, Heald (1971) estimated that about 50% of the aboveground productivity of a mangrove swamp in southwestern Florida was exported to the adjacent estuary as particulate organic matter. From 33 to 60% of the total particulate organic matter in the estuary came from *Rhizophora* (red mangrove) material. The production of organic matter was greater in the summer (the wet season in Florida) than other season, although detrital levels in the swamp waters were greatest from November through February, which is the beginning of the dry season. Thirty percent of the yearly detrital export occurred during November. Heald also found that as the debris decomposed, its protein content increased. The apparent cause of this enrichment, also noted in salt marsh studies, is the increase of bacterial and fungal populations.

Since those early studies, an abundance of studies have been undertaken on the outwelling from mangroves to adjacent estuaries. S. Y. Lee (1995) identified 21 such studies where export of particulate and dissolved organic carbon, nutrients, and even organisms themselves were investigated. Almost all of the studies agree that there is export of particulate organic carbon from mangroves to the adjacent estuary. One study (Rivera-Monroy et al., 1995) did illustrate, contrary to many other studies, that a mangrove forest in Mexico was a sink for inorganic nitrogen and a source of organic nitrogen. A summary of several studies on carbon export from mangrove wetlands gave a range of 2 to 400 g C m^{-2} yr^{-1} with an average of about 200 g C m^{-2} yr^{-1}, about double that exported from salt marshes (Twilley, 1998). In a comparison of leaf litter production and organic export for riverine, fringe, and basin mangrove systems (see Figure 2-21), riverine mangrove systems exported a majority of their organic litter (94%, or 470 g C m^{-2} yr^{-1}), whereas basin mangroves exported much less (21%, or 64 g C m^{-2} yr^{-1}), leaving the leaf litter to decompose or accumulate as peat. The proportion of litterfall production that is exported as well as the total amount of litter that is exported increase as the tidal influence increases.

Effects on the Estuary

The role of mangrove wetlands as both a habitat and a source of food for estuarine fisheries is one of the most often cited functions of these ecosystems. The fact that

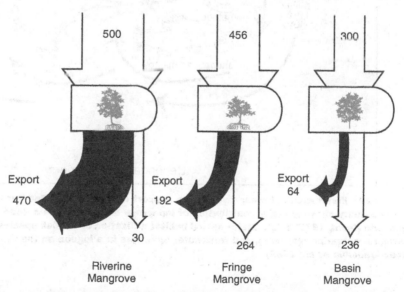

Figure 2-21 Organic carbon fluxes through mangrove swamps: inflows (litterfall), export to adjacent aquatic systems, and other losses (decomposition and peat production). Width of each pathway is proportional to flow (in g C m^{-2} yr^{-1}). (_After Twilley et al., 1986_)

a.

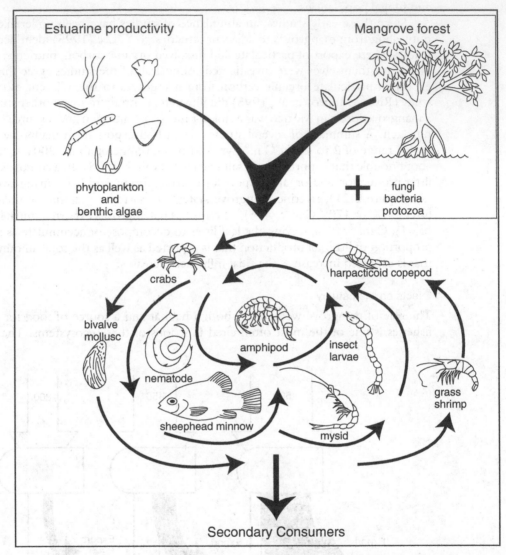

Figure 2-22   Illustrations of mangrove swamp support of fisheries. a. Detritus-based food web in south Florida estuary showing major contribution of mangrove detritus to the estuarine food chain. (*After W. E. Odum and Heald, 1972*) b. Life histories and habitat utilization of six fish species including marine–estuarine spawners, estuarine spawners, and freshwater spawners in a lagoon on the Gulf of Mexico, Mexico. (*After Yánez-Arancibia et al., 1988*)

b.

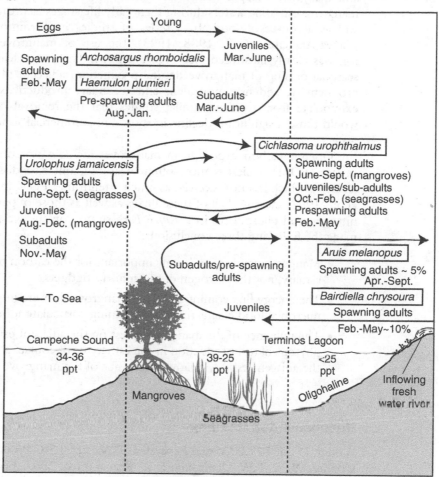

Figure 2-22 (*continued*)

organic carbon is exported from mangrove wetlands does not guarantee that it enters estuarine food chains. Yet, several independent studies have verified that mangrove wetlands are important nursery areas and sources of food for sport and commercial fisheries (see Figure 2-22). Early studies by W. E. Odum (1970) and W. E. Odum and Heald (1972) established that detrital export is important to sport and commercial fisheries (see Figure 2-22a). Through the examination of the stomach contents of more than 80 estuarine animals, Odum found that mangrove detritus, particularly from *Rhizophora*, is the primary food source in the estuary. Important consumers include the spiny lobster (*Panulirus argus*), pink shrimp (*Penaeus duorarum*), mullet (*Mugil cephalus*), tarpon (*Megalops atlanticus*), snook (*Centropomus undecimalis*), and mangrove snapper (*Lutjanus apodus*). The primary consumers also used the mangrove estuarine waters during their early life stages as protection from predators and as a source of food. In a series of studies in Terminos Lagoon in Mexico (Yáñez-Arancibia et al., 1988, 1993), the use of mangrove forests by estuarine fish was clearly illustrated (see Figure 2-22b). There are clear connections among seasonal pulsing of mangrove detrital production, adjacent planktonic and seagrass productivity, and fish movement and secondary productivity. It is reasonable to extrapolate from this and similar studies that the removal of mangrove wetlands would cause a significant decline in sport and commercial fisheries in adjacent open waters.

Organic carbon export from mangroves includes not only particulate organic matter (POM), which is most often measured, but also dissolved organic matter (DOM), which has not received as much attention as POM in export studies. Wafar et al. (1997) measured all of these fluxes as well as those of particulate and dissolved nitrogen and phosphorus in a *Rhizophora* swamp on the west coast of India. They made the following three conclusions:

- Mangrove production is only important for the carbon budget of the adjacent estuary, not the nitrogen or phosphorus budget.
- The energy flux coming from the mangroves is more important for sustaining microbial food chains than for sustaining particulate food chains.
- The influence of the mangrove forest on the adjacent estuary relative to phytoplankton production in the estuary is, of course, a function the size of the adjacent estuary relative to the size of the mangrove forests.

Recommended Readings

Alongi, D. M. 1998. Coastal Ecosystem Processes. CRC Press, Boca Raton, FL.

Conner, W.H., T.W. Doyle, and K.W. Krauss (eds). 2007. Ecology of Tidal Freshwater Forested Wetlands of the Southeastern United States. Springer, Inc., Dordrecht, The Netherlands.

Morris, J. T., P. V. Sundareshwar, C. T. Nietch, B. Kjerfve, and D. R. Cahoon. 2002. Response of coastal wetlands to rising sea level. Ecology 83: 2869–2877.

Rützler, K., and I. C. Feller. 1996. Caribbean mangrove swamps. Scientific American. March 1996: 94–99.

Saenger, P. 2001. Mangrove Ecology, Silviculture and Conservation. Kluwer Academic Publishers, Dordrecht, The Netherlands.

Twilley, R. R. 1998. Mangrove wetlands. In M. G. Messina and W. H. Conner, eds. Southern Forest Wetlands: Ecology and Management. Lewis Publishers, Boca Raton, FL, pp. 445–473.

Weinstein, M. P., and D. A. Kreeger, eds. 2000. Concepts and Controversies in Tidal Marsh Ecology. Kluwer Academic Publishers, Dordrecht, The Netherlands.

Chapter 3

Freshwater Swamps and Marshes

Approximately 90–95% of the world's wetlands are inland or "non-tidal." To simplify matters, we divided these wetlands into freshwater forested swamps and freshwater marshes, which we cover in this chapter, and peatlands, which we cover in the next. Freshwater forested wetlands in North America range from deepwater swamps dominated by bald cypress–tupelo (Taxodium distichum–Nyssa aquatica), *pond cypress–black gum* (Taxodium distichum var. nutans–Nyssa sylvatica var. biflora), *and Atlantic white cedar* (Chamaecyparis thyoides) *swamps found along the eastern seaboard of the United States, to less wet red maple* (Acer rubrum) *swamps found throughout New England and the Mid-Atlantic states and bottomland hardwood forests found along rivers in humid climates everywhere. Trees in these forested wetlands have developed several unique adaptations to the wetland environment, including knees, wide buttresses, adventitious roots, fluted trunks, and gas transport to the rhizosphere. Forested swamp primary productivity is closely tied to hydrologic conditions with lower productivity whenever conditions are either too wet or too dry.*

Inland freshwater marshes are perhaps the most diverse of the wetland types discussed in this book and include the pothole marshes of the north-central United States and south-central Canada, the Florida Everglades, many wetland expanses in the Pantanal in Brazil, and the floodplains of the Okavango Delta in Botswana. Vegetation in freshwater marshes is characterized by tall graminoids such as Typha *and* Phragmites; *the grasses* Panicum *and* Cladium; *the sedges* Scirpus, Schoenoplectus, Cyperus, *and* Carex; *broad-leaved monocots such as* Sagittaria *spp.; and floating aquatic plants such as* Nymphaea *and* Nelumbo. *Some inland marshes such as the prairie pothole marshes follow a 5- to 20-year cycle that includes drought, reflooding, and herbivory. In contrast to bogs, inland marshes have high-pH substrates, high available soil calcium, medium or high loading rates for nutrients, high productivity, and high soil microbial activity that leads to rapid decomposition and recycling, and nitrogen fixation. Peat may or may not accumulate. Most of the primary productivity is routed through detrital pathways, but herbivory can be seasonally important, particularly by muskrats and geese. Inland marshes are valuable as wildlife islands in the middle of agricultural landscapes and have been tested extensively as sites for assimilating nutrients.*

Most of the wetlands of the world are not located along the coastlines but are found in interior regions (these wetlands are called "nontidal" in coastal regions to distinguish them from coastal wetlands; no such term is used by inland wetland scientists). We estimate that there are about 5.5 million km^2 of inland wetlands in the world (see Table 3-1) or about 95% of all the world's wetlands. About 41.5 million ha or about 95% of the total wetlands in the conterminous United States are inland. This estimate includes about 2.6 million ha of non-vegetated freshwater ponds. If Alaska is included, there are 110 million ha of inland wetlands in the United States. It is difficult to put these inland wetlands into simple categories. Our simplified scheme divides them into two groups: freshwater swamps and marshes in this chapter (see Figure 3-1) and peatlands in the next chapter. These divisions roughly parallel the divisions that persist in both the scientific literature and the specializations of wetland scientists.

Terminology for wetlands, especially inland freshwater wetlands, can be confusing and contradictory. In Europe, for example, the term *reed swamp* is often used to describe one type of freshwater marsh dominated by *Phragmites* spp., whereas, in the United States, the word *swamp* usually refers to a forested wetland as described in this chapter. Although the use of classifying terms connotes clear boundaries between different wetland types, in reality they form a continuum. The extremes of freshwater marshes are clearly different, but at the boundaries between two wetland types (e.g.,

Table 3-1   Estimated Area of Inland Wetlands in the World and North America (× 1,000 ha)

	Peatlands	Freshwater Marshes	Freshwater Swamps	Total
World	350,000[a]	95,000[b]	109,000[c]	554,000
Conterminous USA[d]	3,700	9,600	28,200	41,500
Alaska[f]	51,800[e]	17,000	[e]	68,800
Canada[g]	110,000[e]	15,900	[e]	125,900

[a]Bridgham et al., (2001).
[b]Average of several independent estimates.
[c]Matthews and Fung (1987).
[d]Dahl (2006); freshwater marshes includes freshwater emergent and freshwater non-vegetated wetlands.
[e]All palustrine forested wetlands in Alaska and Canada are assumed to be peatlands.
[f]Hall et al. (1994).
[g]Zoltai (1988).

a.

b.

Figure 3-1 Freshwater wetlands discussed in this chapter: a. forested swamp (photo from Florida); and b. freshwater marsh (photo from Ohio). Photos by W.J. Mitsch.

marsh and bog), the distinction is not always clear. *Marshes* (and *reed swamps*, as the term is used in Europe) have mineral soils rather than peat soils. American terminology has developed without much regard to whether the system is peat forming. The fact is that most freshwater marshes and swamps, regardless of where they are located and regardless of their geological origins, accumulate some peat.

Freshwater Swamps

In the nomenclature used in this book, swamps are forested wetlands. We discussed saltwater mangroves and tidal freshwater forested swamps in Chapter 2, and northern peatlands that include trees are discussed in Chapter 4. This section covers the rest of the forested wetlands that are neither tidal nor growing in deep peat; these can be simply described as non-tidal, non-peat-accumulating, palustrine forested wetlands or, more simply, as freshwater swamps. There are an estimated 1.1 million km^2 of freshwater swamps in the world, or about 20% of the inland wetlands of the world (refer to Table 3-1).

Very few trees flourish in standing water. Exceptions are found in the southeastern United States, where cypress (*Taxodium* sp.) and tupelo/gum (*Nyssa* sp.) swamps are found in deepwater forested wetlands and are characterized by bald cypress–water tupelo and pond cypress–black gum communities with permanent or near-permanent standing water. The so-called "deepwater swamp" was defined by Penfound (1952) as having "fresh water, woody communities with water throughout most or all of the growing season." These include cypress domes and alluvial cypress swamps along rivers. Along the middle-eastern seaboard of the United States and along the panhandle of Florida, the cypress swamp partially gives way to another forested wetland, the Atlantic white cedar (*Chamaecyparis thyoides*) swamp. Farther northeast through New England and well into the Midwest, other types of freshwater forested wetlands occur, although they are not as wet as the cypress–tupelo swamps, nor are the tree species coniferous as are cypress and Atlantic white cedar. These broad-leaved deciduous forested wetlands include forests found along river floodplains (riparian forests or bottomland hardwood forests) and a multitude of forested wetlands that are found in isolated upland depressions.

Extensive tracts of riparian wetlands, which occur along rivers and streams, are occasionally flooded by those bodies of water but are otherwise dry for varying portions of the growing season. Riparian forests and freshwater swamps combined constitute the most extensive class of wetlands in the United States, covering an estimated 28 million ha (see Table 3-1). In the southeastern United States, riparian ecosystems are referred to as *bottomland hardwood forests*. They contain diverse vegetation that varies along gradients of flooding frequency. Riparian wetlands also occur in arid and semiarid regions of the United States, where they are often a conspicuous feature of the landscape in contrast to the surrounding arid grasslands and desert. Riparian ecosystems are generally considered to be more productive than the adjacent uplands because of the periodic inflow of nutrients, especially when flooding is seasonal rather than continuous.

Geographical Extent

Cypress–Tupelo Swamps

Bald cypress (*Taxodium distichum* [L.] Rich.) swamps are found as far north as southern Illinois and western Kentucky in the Mississippi River embayment and southern New Jersey along the Atlantic Coastal Plain in the United States (see Figure 3-2a). Pond cypress (*Taxodium distichum* var. *nutans* [Alt.] Sweet), described variously as either a different species or a subspecies of bald cypress, has a more limited range than bald cypress and is found primarily in Florida and southern Georgia; it is not present along the Mississippi River floodplain except in southeastern Louisiana. Another species indicative of the deepwater swamp is the water tupelo (*Nyssa aquatica* L.), which has a range similar to that of bald cypress along the Atlantic Coastal Plain and the Mississippi River, although it is generally absent from Florida except for the western peninsula. Water tupelo occurs in pure stands or is mixed with bald cypress in floodplain swamps.

White Cedar Swamps

White cedar swamps, dominated by Atlantic white cedar (*Chamaecyparis thyoides* [L.] BSP), are found along the Atlantic and Gulf of Mexico coastlines of the United States as far north as southeastern Maine (see Figure 3-2b). These wetlands are not nearly as plentiful as are cypress–tupelo swamps. White cedar occurs in about 215,000 ha of forestland, but the species accounts for a majority of the trees in only about 44,200 ha (Sheffield et al., 1998). Only 5,300 ha of Atlantic white cedar swamps remain in the glaciated northeastern United States, with red maple swamps now more prevalent there. The three states that have the most area of timberland with Atlantic white cedar are North Carolina, Florida, and New Jersey. The regions with the highest concentrations of Atlantic white cedar are the Pinelands of southeastern New Jersey; the Dismal Swamp of Virginia and North Carolina; and the floodplains of the Escambia, Apalachicola, and Blackwater Rivers in Florida (Sheffield et al., 1998).

Red Maple Swamps

One of the most common of the broad-leaved deciduous forested wetlands in the northeastern United States is the red maple (*Acer rubrum* L.) swamp. Toward the west into Pennsylvania and Ohio, red maple swamps are replaced by swamps dominated by trees such as ash (*Fraxinus* spp.), American elm (*Ulmus americana*), swamp white oak (*Quercus bicolor*), and a number of other species, but in the northeastern United States, the red maple swamp is the most common swamp. Using an approximation that all broad-leaved deciduous forested wetlands in several of the coastal states of the northeastern United States are red maple swamps (this approximation would not apply west or south of New York), Golet et al. (1993) estimated that there are 353,000 ha of red maple swamps in these six states. Red maple forests also occur in the Upper Peninsula of Michigan and northeastern Wisconsin. The range of the species *Acer rubrum* extends westward to the Mississippi River and northward through much of Ontario and parts of Manitoba and Newfoundland, but the tree can grow in both

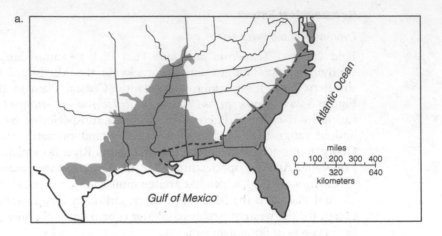

Figure 3-2 **Distribution of dominant forested wetland trees in southeastern United States: a. bald cypress (*Taxodium distichum*) and pond cypress (*Taxodium distichum* var. *nutans*) (with dotted line indicating northern extent of pond cypress) and b. white cedar (*Chamaecyparis thyoldes*). (*After Little, 1971; Laderman, 1989*)**

wetlands and dry, sandy or rocky uplands. Thus, the presence of red maple does not always indicate wetlands, as would the presence of cypress, tupelo, or white cedar.

Riparian Ecosystems

Riparian ecosystems are not as easily defined by location. Abernethy and Turner (1987) estimated 22.9 million ha of riparian forests of the United States, excluding California, Arizona, New Mexico, Hawaii, and Alaska. Forested wetland areas in Arizona, California, and New Mexico total about 360,000 ha (Brinson et al., 1981), and 12 million ha are estimated to be in Alaska. Between 1940 and 1980 the national loss rate was 0.27 percent per year, or about 2.8 million ha. Most of this loss occurred in the south-central and southeastern United States, which has 58 percent of the total United States forested wetlands. That loss rate has slowed considerably since then.

Mesic riparian ecosystems, commonly called *bottomland hardwood forests* or bottomland hardwoods, are one of the dominant types of riparian ecosystems in the United States. Historically the term bottomland hardwood forest has been used to describe the vast forests that occur on river floodplains of the eastern and central United States, especially in the Southeast. Bottomland hardwood forests are particularly notable wetlands because of the large areas that they cover in the southeastern United States and because of the rapid rate at which they are being converted to other uses such as agriculture and human settlements. This ecosystem is particularly prevalent in the lower Mississippi River alluvial valley as far north as southern Illinois and western Kentucky (Taylor et al., 1990) and along many streams that drain into the Atlantic Ocean on the south Atlantic Coastal Plain. The Nature Conservancy (1992) estimated that before European settlement the Mississippi River alluvial plain supported about 21 million ha of riparian forests; about 4.9 million ha remained as of 1991. The Atlantic Coastal Plain from Maryland to Florida is another area of dense riparian forests lining the many rivers that flow into the ocean. These forests have been logged, as have most in this country, but many are otherwise fairly intact.

In contrast to the broad, flat, expansive southeastern riparian forests, western United States riparian zones tend to be narrow, linear features of the landscape, often lining streams with steep gradients and narrow floodplains. Along southeastern high-order rivers the contrast in elevation and vegetation between bottomland and upland is often subtle and the gradients are gradual, whereas in the West gradients are usually sharp and the visual distinctions are usually clear. Western riparian zones have been extensively modified by human activity. Conversion to housing or agriculture is widespread. Damage from grazing animals is almost ubiquitous. In an area where vegetation is generally limited by the lack of water, riparian vegetation and the availability of water inevitably draw and concentrate cattle. Grazing along these primarily low-order streams results in increased erosion and channel downcutting while higher-order streams have been modified for water use.

Geomorphology and Hydrology

Cypress Swamps

Southern cypress–tupelo swamps occur under a variety of geologic and hydrologic conditions, ranging from the extremely nutrient-poor dwarf cypress communities of southern Florida to the rich floodplain swamps along many tributaries of the lower Mississippi River basin. A useful classification of deepwater swamps in terms of their geological and hydrological conditions includes the following types (see Figure 3-3):

- **Cypress domes:** Cypress domes (sometimes called cypress ponds or cypress heads) are poorly drained to permanently wet depressions dominated by pond cypress. They are generally small in size, usually 1 to 10 ha, and are numerous in the upland pine flatwoods of Florida and southern Georgia. Cypress domes are found in both sandy and clay soils and usually have several centimeters of organic matter that has accumulated in the wetland depression. These wetlands are called *domes* because of their appearance when viewed from the side: The larger trees are in the middle, and smaller trees are toward the edges (see Figure 3-3a). Ewel and Wickenheiser (1988) confirmed that trees grow slowest at the edges and fastest near the center of the domes but found no significant differences in tree growth among small, medium, and large cypress domes. This "dome" phenomenon, it has been suggested, is caused by deeper peat deposits in the middle of the dome, fire that is more frequent around the edges of the dome, or a gradual increase in the water level that causes the dome to "grow" from the center outward (Vernon, 1947; Kurz and Wagner, 1953). A definite reason for this profile has not been determined, nor do all domes display the characteristic shape.

- **Dwarf cypress swamps:** There are major areas in southwestern Florida, primarily in the Big Cypress Swamp and the Everglades, where pond cypress is the dominant tree but it grows stunted and scattered in a herbaceous understory marsh (see Figure 3-3b). The trees generally do not grow more than 6 or 7 m high and are more typically 3 m in height. These wetlands are called dwarf cypress or pigmy cypress swamps. The poor growing conditions are caused primarily by the lack of suitable substrate overlying the bedrock limestone that is found in outcrops throughout the region. The hydroperiod includes a relatively short period of flooding as compared with other deepwater swamps, and fire often occurs. The cypress, however, are rarely killed by fire because of the lack of litter accumulation and buildup of fuel.

- **Lake-edge swamps:** Bald cypress swamps are also found as margins around many lakes and isolated sloughs in southeastern United States, ranging from Florida to southern Illinois (see Figure 3-3c). Tupelo and water-tolerant hardwoods such as ash (*Fraxinus* spp.) often grow in association with the bald cypress. A seasonally fluctuating water level is characteristic of these systems and is necessary for seedling survival. The trees in these systems receive nutrients from the lake as well as from upland runoff. The lake-edge swamp has been described as a filter that receives overland flow from the uplands and allows sediments to settle out and chemicals to adsorb onto the sediments before the water discharges into the open lake. The importance of this filtering function, however, has not been adequately investigated.

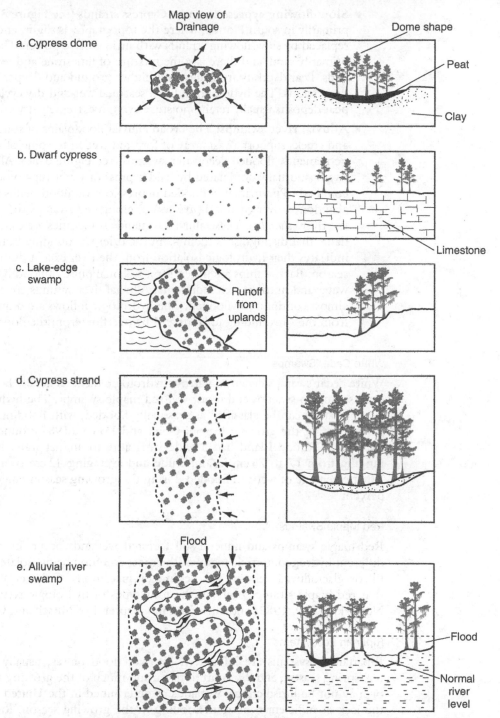

Map view of Drainage

a. Cypress dome

Dome shape

Peat

Clay

b. Dwarf cypress

Limestone

c. Lake-edge swamp

Runoff from uplands

d. Cypress strand

e. Alluvial river swamp

Flood

Flood

Normal river level

Figure 3-3 General profile and flow pattern of major types of deepwater swamps, showing a. cypress dome, b. dwarf cypress, c. lake-edge swamp, d. cypress strand, and e. alluvial river swamp. (*After H. T. Odum, 1982*)

- **Slow-flowing cypress strands:** Cypress strands (see Figure 3-3d) are found primarily in south Florida, where the topography is slight, and rivers are replaced by slow-flowing strands with little erosive power. The substrate is primarily sand, and there is some mixture of limestone and remnants of shell beds. Peat deposits are shallow on higher ground and deeper in the depressions. The hydroperiod has a seasonal wet-and-dry cycle. The deeper peat deposits usually retain moisture even in extremely dry conditions.

- **Alluvial river swamps:** The broad alluvial floodplains of southeastern rivers and creeks support a vast array of forested wetlands, some of which are permanently flooded deepwater swamps (see Figure 3-3e). Alluvial river swamps, usually dominated by bald cypress or water tupelo or both, are confined to permanently flooded depressions on flood-plains such as abandoned river channels (*oxbows*) or elongated swamps that usually parallel the river (*sloughs*). These alluvial wetlands sometimes are called *backswamps*, a name that distinguishes them from the drier surrounding bottomlands and indicates their hydrologic isolation from the river except during the flooding season. Backswamps are noted for a seasonal pulse of flooding that brings in water and nutrient-rich sediments. Alluvial river swamps are continuously or almost continuously flooded. The hydrologic inflows are dominated by runoff from the surrounding uplands and by overflow from the flooding rivers.

White Cedar Swamps

White cedar swamps occupy a narrow hydrologic niche generally between deepwater cypress–tupelo swamps and moist-soil red maple swamps. The hydrologic regime of cedar swamps can be classified as seasonally flooded, with flooding for an extended period during the growing season. Golet and Lowry (1987) found that a group of swamps in Rhode Island had a wide variability in annual water-level fluctuations, ranging from 17 to 75 cm in amplitude and averaging 42 cm over a 7-year period. The percentage of wetland flooded during the growing season ranged from 18 to 76 percent.

Red Maple Swamps

Red maple swamps and mineral soil forested wetlands occur, in general, in several different hydrogeomorphic regimes, the most common being isolated basins in glacial till or glaciofluvial deposits left behind by glaciations. Golet et al. (1993) suggest that red maple swamps occur in the four types of hydrologic settings described by Novitzki (1979, 1982) and illustrated in Chapter 4 of Mitsch and Gosselink (2007).

Riparian Ecosystems

Riparian ecosystems are influenced by river flood pulses, usually in the wet winter/spring season, and dry conditions during much of the growing season. They may or may not be jurisdictional wetlands as determined in the United States because of the lack of sufficient root-zone flooding in the growing season. Riparian vegetation along a stream or river is determined by the cross-sectional morphology, including braiding of the stream, width of the floodplain, soil type, and elevation and moisture gradients. These are all determined in part by larger scale (continental, basin, stream system) processes that are modified by local biotic and physical processes. The riparian soil moisture regime, in large part, explains the plant associations. The relationship, however, is seldom that simple. Soil moisture and depth to groundwater are not the only factors governing plant establishment. Low floodplain elevations are often swept clean of plants by floods, so that seedlings do not survive, and the vegetation is limited to annuals and perennials that survive until the next flood. Trees mature only at elevations above moderate floods where they can become well enough established to withstand the severe floods. The extensive bottomland hardwood forests of the southeastern United States show clear zonation patterns related to elevation and flood frequency. In these broad floodplains elevations are not monotonic gradients, because the floodplain level does not necessarily rise uniformly from the river.

Biogeochemistry

Lockaby and Walbridge (1998) describe the biogeochemistry of forested wetlands as "the most complex and difficult to study with any forest ecosystem type." Forested wetlands have soil and water chemistry that varies from the rich sediments of alluvial cypress swamps to the extremely low mineral and acidic waters of surface water depression red maple swamps and cypress domes. Wide ranges of pH, dissolved substances, and nutrients are found in the soils and waters of these swamps. Several facts should be noted from this wide range of soil and water chemistry:

- Swamps are generally acidic to circum-neutral, depending on the accumulation of peat and the degree to which precipitation dominates the hydrology.
- Nutrient conditions vary from nutrient- and mineral-poor conditions in rainwater-fed swamps to nutrient- and mineral-rich conditions in alluvial river swamps and groundwater discharge swamps.
- An alluvial river swamp often has water quality very different from that of the adjacent river. Swamps in alluvial settings are generally fed by both groundwater discharge and flooding rivers and can have water chemistry quite different from either source.

Many freshwater swamps, particularly alluvial river swamps, are "open" to river flooding and other inputs of neutral and generally well-mineralized waters. The pH of many alluvial swamps in the southeastern United States is 6 to 7, and there are high concentrations of dissolved ions. Cypress domes and perched-basin swamps, on the other hand, are fed primarily by rainwater and have acidic waters, usually in the pH range of 3.5 to 5.0, caused by humic acids produced within the swamp. Colloidal humic substances contribute to both the low pH and the tea-colored or "blackwater" appearance of the standing water in many forested wetlands.

The buffering capacity of the water in isolated swamps such as cypress domes is low. Little or no alkalinity and low concentrations of dissolved ions and nutrients are characteristic of precipitation-dominated cypress domes and some Atlantic white cedar and red maple swamps. These wetlands have much in common with the oligotrophic or ombrotrophic peatlands described in the next chapter. On the other hand, swamps open to major surface water and groundwater inputs are generally rich in alkalinity, dissolved ions, and nutrients. For example, conductivity of surface water ranges from only 60 μS/cm in cypress domes in Florida to 200 to 400 μS/cm in alluvial cypress swamps in Kentucky and Illinois. Cypress domes and dwarf cypress swamps are low in nutrients because of their relative hydrologic isolation. Nutrients are often more plentiful in alluvial river swamps, particularly during flooding from the adjacent river.

In an extensive literature review, Bedford et al. (1999) found that temperate freshwater swamps in North America tend to have an N:P soil ratio <14 (suggesting possible N-limitation) and noted that N and P availability are probably dictated by trophic condition. For instance, Schilling and Lockaby (2006) found strong evidence that the aboveground productivity of floodplain swamps along the Satilla River (an oligotrophic blackwater river in Georgia) was P-limited based on high nutrient use efficiency, C:P ratios, and P:lignin ratios of litterfall. These findings were in contrast to swamps along the Altamaha River (a eutrophic redwater river in Georgia), which appeared to be more N-limited. Aside from N and P, there was also evidence in the Satilla River swamps that certain base cations (Ca and K) may also limit floodplain swamp productivity.

In a comparison of an Atlantic white cedar swamp with adjacent forested wetlands and nonforested peatlands, Whigham and Richardson (1988) found cedar swamp soils to be significantly higher in pH, calcium, and magnesium than the other sites (see Table 3-2), suggesting a groundwater or brackish-water source might be important for Atlantic white cedar to compete with other swamp trees. Phosphorus was lowest in the white cedar swamp, suggesting this was the most significant limiting nutrient. The high pH measured in this study suggests that Atlantic white cedar may do best

Table 3-2 Soil Chemistry of *Chamaecyparis thyoides* and *Acer rubrum* Swamps in Maryland Compared to Nonforested Peatlands

Soil Parameters (top 50 cm)	White Cedar Swamp	Red Maple Swamp	Nonforested Peatlands
pH	5.34	4.23	4.54
Organic matter (%)	59±5	67±3	68±2
Nitrogen (%)	1.6±0.1	1.5±0.1	1.7±0.1
Phosphorus (%)	0.07±0.01	0.24±0.03	0.10±0.01
NO_3–N (μg/g)	0.8±0.1	0.3±0.1	0.5±0.1
NH_4–N (μg/g)	67±4	72±19	76±10
Ca^{2+} (μg/g)	1,810	339	710
Mg^{2+} (μg/g)	1,420	493	477
K^+ (μg/g)	1,054	1,622	857
Na^+ (μg/g)	841	134	383
Fe (mg/g)	6.3	5.9	5.4
Al (mg/g)	8.0	5.4	7.6

Source: Adapted from Whigham and Richardson (1988).

in sites with high pH, although these swamps have been reported to occur under low-pH (3.2–4.4) conditions in the Great Dismal Swamp (F. Day, 1984).

In riparian forest soils, phosphorus availability has been shown to increase during floods although the exact reason for this can vary. Wright et al. (2001) examined the availability of P after experimentally flooding plots in a Georgia floodplain forest and found that flooding did release P; however they found that there was no change in Fe/Al phosphate factions. The reduction of Fe (III) phosphates and hydrolysis of Al phosphates has often been credited for increased P availability after soils have been flooded. While this may occur when upland soils are flooded, the authors attributed the increased available P during floods to biological processes such the release of P from microbial biomass and the suppression of biological P demand during anaerobic conditions.

Vegetation

Cypress Swamps

Southern deepwater swamps, particularly cypress wetlands, have plant communities that either depend on or adapt to the almost continuously wet environment. There are several distinctions between bald cypress and pond cypress swamps. The dominant canopy vegetation found in alluvial river swamps of the southeastern United States includes bald cypress (*Taxodium distichum*) and water tupelo (*Nyssa aquatica*). The trees are often found growing in association in the same swamp, although pure stands of either bald cypress or water tupelo are also frequent in the southeastern United States. Many of the pure tupelo stands may have been the result of the selective logging of bald cypress. The pond cypress-black gum (*Taxodium distichum* var. *nutans–Nyssa sylvatica* var. *biflora* [Walt.] Sarg.) swamp is more commonly found on the uplands of the southeastern Coastal Plain, usually in areas of poor sandy soils without alluvial flooding (see Table 3-3). These same conditions are usually found in cypress domes.

One of the main features that distinguishes bald cypress trees from pond cypress trees is the leaf structure (see Figure 3-4). Bald cypress has needles that spread from the twig in a flat plane, whereas pond cypress needles are appressed to the twig. Both species are intolerant of salt and are found only in freshwater areas. Pond cypress is limited to sites that are poor in nutrients and are relatively isolated from the effects of river flooding or large inflows of nutrients.

When deepwater swamps are drained or when their dry period is extended dramatically, they can be invaded by pine (e.g., *Pinus elliottii*) or hardwood species. In north-central Florida, a cypress–pine association indicates a drained cypress dome (Mitsch and Ewel, 1979). Hardwoods that characteristically are found in cypress domes include swamp red bay (*Persea palustris*) and sweet bay (*Magnolia virginiana*). In lake-edge and alluvial river swamps, several species of ash (*Fraxinus* sp.) and maple (*Acer* sp.) often grow as subdominants with the cypress or tupelo or both. In the Deep South, Spanish moss (*Tillandsia usneoides*) is found in abundance as an epiphyte on the stems and branches of the canopy trees.

Table 3-3 Distinction between Bald Cypress and Pond Cypress Swamps

Characteristic	Bald Cypress Swamp	Pond Cypress Swamp
Dominant cypress	*Taxodium distichum*	*Taxodium distichum* var. *nutans*
Dominant tupelo or gum (when present)	*Nyssa aquatica* (water tupelo)	*Nyssa sylvatica* var. *biflora* (black gum)
Tree physiology	Large, old trees, high growth rate, usually abundance of knees and spreading buttresses	Smaller, younger trees, low growth rate, some knees and buttresses but not as pronounced
Location	Alluvial floodplains of Coastal Plain, particularly along Atlantic seaboard, Gulf seaboard, and Mississippi embayment	"Uplands" of Coastal Plain, particularly in Florida and southern Georgia
Chemical status	Neutral to slightly acidic, high in dissolved ions, usually high in suspended sediments and rich in nutrients	Low pH, poorly buffered, low in dissolved ions, poor in nutrients
Annual flooding from river	Yes	No
Types of deepwater swamps	Alluvial river swamp, cypress strand, lake-edge swamp	Cypress dome, dwarf cypress swamp

The abundance of understory vegetation in cypress–tupelo swamps depends on the amount of light that penetrates the tree canopy. Many mature swamps appear as quiet, dark cathedrals of tree trunks devoid of any understory vegetation. Even when enough light is available for understory vegetation, it is difficult to generalize about its composition. There can be a dominance of woody shrubs, of herbaceous vegetation, or of both. Fetterbush (*Lyonia lucida*), wax myrtle (*Myrica cerifera*), and Virginia willow (*Itea virginica*) are common as shrubs and small trees in nutrient-poor cypress domes. Understory species in higher nutrient river swamps include buttonbush (*Cephalanthus occidentalis*) and Virginia willow. Some continually flooded cypress swamps that have high concentrations of dissolved nutrients in the water develop dense mats of duckweed (e.g., *Lemna* spp., or *Spirodela* spp., or *Azolla* spp.) on the water surface during most of the year. Floating logs and old tree stumps often provide substrate for understory vegetation to attach and to flourish.

White Cedar Swamps

Cedar swamps occur within a wide climatic range along the East Coast of the United States and in an intermediate hydrology between deepwater cypress swamps in the South and forested swamps such as red maple swamps in the North. Often thought of as monospecific, even-aged stands with tightly spaced *Chamaecyparis thyoides* trees, cedar swamps in these cases would have no subcanopy, few shrubs, and little in the herbaceous layer. However, the tree is often found in mixed stands, with co-dominants such as *Betula populifolia* (gray birch), *Picea mariana* (black spruce), *Pinus strobus* (Eastern white pine), and *Tsuga canadensis* (Eastern hemlock) (Laderman, 1989). In the South, co-dominant trees include *Gordonia lasianthus* (loblolly bay), *Persea*

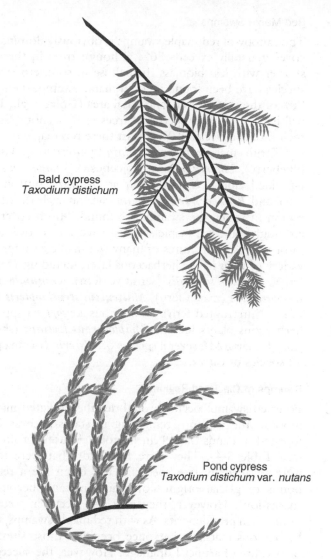

Figure 3-4   Distinction of leaves between top: bald cypress (*Taxodium distichum*) and bottom: pond cypress (*Taxodium distichum* var. *nutans*).

borbonia (red bay), *P. palustris* (swamp red bay), and *Taxodium districhum* (bald cypress).

The shrub layer in cedar swamps with relatively open canopies includes many ericaceous shrubs such as *Aronia arbutifolia* (red chokeberry), *Clethra alnifolia* (sweet pepperbush), *Ilex glabra* (gallberry), *Leucothoe racemosa* (fetterbush), and *Vaccinium corymbosum* (highbush blueberry) (Laderman, 1989).

Red Maple Swamps

The canopy of red maple swamps is obviously dominated by *Acer rubrum* L. Canopy cover generally exceeds 80%, although trees in these northern swamps tend to be shorter with less biomass than those in southern swamps. Although up to 50 tree species have been found in a red maple swamp, the red maple can account for up to 90% of the stem density and basal area (Golet et al., 1993). In general, a specific site will have about four species of trees in the canopy/subcanopy, depending on which region of the glaciated Northeast these red maple swamps occur.

Shrubs include *Ilex verticillata* (winterberry), *Vaccinium corymbosum* (highbush blueberry), *Lindera benzoin* (spicebush), *Viburnum* spp. (arrowwood), *Alnus rugosa* (speckled alder), *Cephalanthus occidentalis* (buttonbush), *Corylus cornuta* (hazelnut), and *Rhododendron viscosum* (swamp azalea), with dominance depending on the region in which the swamps are found. Shrub cover is generally greater than 50%, although some red maple swamps have shrub cover as low as 6 percent. One of the most interesting features of many red maple swamps is the predominance of a great variety of ferns in the herbaceous layer, including *Osmunda cinnamomea* (cinnamon fern), *Onoclea sensibilis* (sensitive fern), *Osmunda regalis* (royal fern), *Thelypteris thelypteroides* (marsh fern), *Matteuccia struthiopteris* (ostrich fern), *Osmunda claytoniana* (interrupted fern), and various *Dryopteris* spp. (wood ferns). Other common herbaceous plants include *Symplocarpus foetidus* (skunk cabbage), *Caltha palustris* (marsh marigold), several species of *Glyceria* (manna grass), and several of more than 32 species of *Carex*.

Swamps of Glaciated Regions

Forested swamps occur throughout the glaciated midwestern United States; in fact, most of the wetlands remaining in states such as Ohio, Indiana, and Illinois are forested wetlands that occur in isolated basins or floodplains amid agricultural fields (see Table 3-4). They were the fields that were too wet to plant and gradually were invaded by tree species. They often are a remnant of a gradual process of ponds of glacial origin slowly infilling and becoming forested (a true hydrarch succession). However, they may also occur in wet basins on mineral hydric soils rather than peat deposits. As with red maple swamps, the trees generally have replaced herbaceous marshes that once occupied those sites, because of natural succession or because of artificial drainage. However, the succession of these systems is poorly understood.

Fire in Swamps?

Fire is generally infrequent in swamps because of standing water or saturated soil conditions, but it can be a significant ecological factor during droughts or in swamps that have been artificially drained. In general, fire is more frequent in the forested swamps of Florida than anywhere else, because of the more frequent lightning storms and because of a predictable dry season. For example, from 1970 to 1977, there were four fires in the Big Cypress National Preserve in southern Florida, each affecting an average of

Table 3-4   Typical Vegetation in a Hardwood Swamp Forest in Ohio[a]

	Wetland Indicator Status[b]
Trees	
Quercus palustrus (pin oak)	FACW
Quercus bicolor (swamp white oak)	FACW
Acer saccharinum (silver maple)	FACW
Acer rubrum (red maple)	FAC
Ulmus americana (American elm)	FACW
Fraxinus pennsylvanica (green ash)	FACW
Shrubs/Understory	
Lindera benzoin (spicebush)	FACW
Cephalanthus occidentalis (buttonbush)	OBL
Rosa multiflora[c] (multiflora rose)	FACU
Carpinus caroliniana (hornbeam, ironwood)	FAC
Herbs	
Polygonum spp. (smartweed)	FAC/OBL
Symplocarpus foetidus (skunk cabbage)	OBL
Lemna spp. (duckweed)	OBL
Alisma plantago-aquatica (water plantain)	OBL
Aster spp. (asters)	FAC/FACW
Carex spp. (sedges)	FACW/OBL
Ranunculus septentrionalis (swamp buttercup)	OBL
Saxifraga pennsylvanica (swamp saxifrage)	OBL
Onoclea sensibilis (sensitive fern)	FACW
Bidens comosa (leafy-bracted beggar-ticks)	FACW
Bidens frondosa (devil's beggar-ticks)	FACW
Scirpus atrovirens (green bulrush)	OBL
Scirpus cyperinus (wool grass)	FACW

[a]Wetland species at Gahanna Woods Nature Preserve, Franklin County, Ohio.
[b]Use of wetland indicator status for the northeastern United States. In order of wet to dry: OBL, obligate wetland plant; FACW, facultative wet plant; FAC, facultative plant; FACU, facultative upland plant.
[c]Nonnative species.

(continued)

500 ha. It appears that fire is rare in most alluvial river swamps but can be more frequent in cypress domes or dwarf cypress swamps—as frequent as several times per century. Fire had a "cleansing" effect on the trees in a cypress dome in north-central Florida, selectively killing almost all of the pine and hardwoods but relatively few pond cypress (Ewel and Mitsch,

(continues)

(*continued*)

1978). This suggests a possible advantage of fire to some shallow cypress ecosystems in eliminating competition that is less water tolerant. Casey and Ewel (2006) identified fire severity as a key factor influencing tree succession in Florida pond cypress swamps. In their generalized succession model, the exclusion of fire (due to geomorphic conditions) tends to promote mixed bay-cypress communities, while periodic moderate fires tend to promote monotypic cypress or cypress-tupelo forests. Severe fires can lead to shrub or marsh conditions.

Fire can also be an influential factor on white cedar swamps. If water is low, fire can be quite destructive, killing cedar trees and burning the peat deeply. If water levels are high, light fire can have a cleansing effect, eliminating shrubs and brush and favoring cedar seedling germination (Laderman, 1989). In *C. thyoides* swamps of the Atlantic Coast, the highly flammable cedar foliage burned frequently (five fires per each 100- to 200-year interval) during pre-European settlement time; when fires became more rare after European settlement, stands of cedar became the familiar dense mono-specific systems that are common today (Motzkin et al., 1993).

Tree Adaptations

Vascular plants, particularly trees, have a difficult time surviving under continuously flooded conditions. Only a handful of species of trees in North America can stay viable in continuous flooding and, even then, their growth is generally slowed; trees that are found in freshwater swamps are stressed with the wet conditions but have found ways to adapt (Mitsch and Gosselink, 2007). The most conspicuous adaptations specific to the major tree species in forested swamps are discussed here.

Knees and Pneumatophores Cypress (bald and pond), water tupelo, and black gum are among a number of wetland plants that produce pneumatophores. In deepwater swamps, these organs extend from the root system to well above the average water level (see Figure 3-5a). On cypress, these "knees" are conical and typically less than 1 m in height, although some cypress knees are as tall as 3 to 4 m. Knees are generally much more prominent on cypress than on tupelo. Pneumatophores on black gum in cypress domes are actually arching or "kinked" roots that approximate the appearance of cypress knees. The functions of the knees have been speculated about for more than a century. It was thought that the knee might be an adaptation for anchoring the tree because of the appearance of a secondary root system beneath the knee that is similar to and smaller than the main root system of the tree. Observations of swamp and upland damage in South Carolina following Hurricane Hugo in 1990 showed that cypress trees often remained standing while hardwoods and pines did not, supporting the tree-anchoring theory for cypress root, knee, and buttress systems (K. Ewel, personal communication).

Figure 3-5 Among the several features of vegetation in cypress swamps are a. cypress knees and b. buttresses and large size and long life of cypress trees.

Other discussions of cypress knee function have centered on their possible use as sites of gas exchange for the root systems. Penfound (1952) argued that cypress knees are often absent where they are most needed—in deep water—and that the wood of the cypress knee is not aerenchymous; that is, there are no intercellular gas spaces capable of transporting oxygen to the root system. However, gas exchange does occur at the knees. S. L. Brown (1981) estimated that gas evolution from knees accounted for 0.04 to 0.12 g C m^{-2} yr^{-1} of the respiration in a cypress dome, and 0.23 g C m^{-2} yr^{-1} in an alluvial river swamp. This accounted for 0.3 to 0.9 percent of the total tree respiration, but 5 to 15 percent of the estimated woody tissue (stems and knees) respiration. The fact that CO_2 is exchanged at the knee, however, does not prove that oxygen transport is taking place there or that the CO_2 evolved was the result of oxidation of anaerobically produced organic compounds in the root system.

Buttresses *Taxodium* and *Nyssa* species and, to a lesser degree, *Chamaecyparis thyoides* often produce swollen bases or buttresses (stem hypertrophy) when they grow in flooded conditions (see Figure 3-5b). The basal swelling can extend from less than 1 m above the soil to several meters, depending on the hydroperiod of the wetland. Swelling generally occurs along the part of the tree that is flooded at least seasonally, although the duration and frequency of the flooding necessary to cause the swelling are unknown. One theory is that the height of the buttress is a response to aeration: The greatest swelling occurs where there is a continual wetting and soaking of the tree trunk but where the trunk is also above the normal water level (Kurz and Demaree, 1934). The value of the buttress swelling to ecosystem survivability is unknown; it may simply be a relict response that is of little use to the plant.

Seed Germination and Dispersal The seeds of swamp trees require oxygen for germination. For example, cypress seeds and seedlings require moist but not flooded soil for germination and survival. Occasional draw-downs, if only at relatively infrequent intervals, are therefore necessary for the survival of trees in these swamps unless floating mats develop. Otherwise, continuous flooding will ultimately lead to an open-water pond.

The dispersal and survival of the seeds of many swamp trees depend on hydrologic conditions. Schneider and Sharitz (1986) found a relatively low number of viable seeds in a seed bank study of a cypress–tupelo swamp in South Carolina. An average of 127 seeds/m^2 were found for woody species (88 percent as cypress or tupelo) in the swamp compared to a seed density of 233 seeds/m^2 from an adjacent bottomland hardwood forest. They speculated that the continual flooding in the cypress–tupelo swamp leads to reduced seed viability. Huenneke and Sharitz (1986) and Schneider and Sharitz (1988) elaborated further on the importance of *hydrochory* (seed dispersal by water) in these swamps. Hydrologic conditions, particularly scouring by flooding waters, are important factors in determining the composition, dispersal, and survival of seeds in riverine settings. Seeds are transported relatively long distances; Schneider and Sharitz (1988) reported that a distance of 1.8 km was necessary in their study to trap 90 percent of seeds from a given tree. They found that flowing water distributes seeds non-randomly: The highest seed densities accumulate near obstructions such as logs, tree stumps, cypress knees, and tree stems, and the lowest seed densities in open-water areas. They also found higher densities of seeds near the edge of the swamps (415 seeds/m^2) than near the center (175 seeds/m^2). Bald cypress and tupelo seeds are produced mainly in the fall and winter, between the periods of both lowest (October) and highest (March) stream flow, giving the fallen seeds the widest possible range of hydrologic conditions. Schneider and Sharitz (1988) concluded that "seed dispersal processes of many wetland species are sensitively linked to the timing and magnitude of hydrologic events."

Longevity

Some swamp trees may live for centuries and achieve great sizes (see Figure 3-5b). One individual cypress in Corkscrew Swamp in southwestern Florida was determined to be about 700 years old; Laderman (1998) reports that the maximum age of *Taxodium* is 1,000 years. By contrast, *Chamaecyparis thyoides* lives to a maximum of 300 years (Clewell and Ward, 1987). Mature bald cypress trees are typically 30 to 40 m in height and 1 to 1.5 m in diameter. Anderson and White (1970) reported a very large cypress tree in a cypress–tupelo swamp in southern Illinois that measured 2.1 m in diameter. C. A. Brown (1984) summarized several reports that documented bald cypress as large as 3.6 to 5.1 m in diameter.

Shallow or Adventitious Roots   Some species, such as *Acer rubrum*, develop very shallow root systems in response to flooding, in all likelihood because the surface soil is closest to the atmospheric source of oxygen. In aerated soils, the same species will develop deep roots. Other swamp species such as willows (*Salix* sp.), green ash (*Fraxinus pennsylvanica*), and cotton-woods (*Populus deltoides*) develop adventitious roots above ground from the stem in response to flooding.

Gaseous Diffusion   Woody trees have a particular problem getting oxygen to their rhizosphere when they are flooded, and few species do it well enough to survive continual flooding. The swamp trees, including *Taxodium, Nyssa, Alnus, Fraxinus*, among others, have the ability to supply oxygen to their root systems in amounts adequate for rhizospheric demands. Solar radiation, which heats up the tree stems by a couple of degrees, causes a light-induced gas flow that can be considerably greater in selected swamp seedlings than in the same trees in the dark: this thermally induced flow of air through vascular plants is called *thermo-osmosis* by some (Grosse et al., 1998), and is enhanced by the development of aerenchymous stem and root tissue (see Mitsch and Gosselink, 2007, Chapter 6).

Consumers

Invertebrates

Invertebrate communities, particularly benthic macroinvertebrates, have been analyzed in several cypress–tupelo swamps. A wide diversity and high number of invertebrates have been found in permanently flooded swamps. Species include crayfish, clams, oligochaete worms, snails, freshwater shrimp, midges, amphipods, and various immature insects. Batzer and Wissinger (1996) reported that insects, particularly midges, can dominate forested wetlands and that midges are most likely to reach high densities. Many of these invertebrates are highly dependent, either directly or indirectly, on the abundant detritus found in these systems. Beck (1977) found that a cypress–tupelo swamp in the Louisiana Atchafalaya Basin had a higher number of invertebrate organisms (3,768 individuals/m^2) than the bayous (3,292/m^2), lakes (1,840/m^2), canals (1,593/m^2), and rivers (327/m^2). These high densities were attributed to the abundance of detritus and to the pulse of spring flooding. Sklar and Conner (1979), working in contiguous environments in Louisiana, found a higher number of benthic organisms in a cypress–tupelo swamp (7,500 individuals/m^2) than in a nearby impounded swamp (3,000/m^2) or in a swamp managed as a crayfish farm. Their study suggests that the natural swamp hydroperiod results in the highest number of benthic invertebrates.

Ziser (1978) and Bryan et al. (1976) surveyed the benthic fauna of alluvial river swamps. Oligochaetes and midges (Chironomidae), both of which can tolerate low-dissolved-oxygen conditions, and amphipods such as *Hyalella azteca*, which occur in abundance amid aquatic plants such as duckweed, usually dominate. In nutrient-poor cypress domes, the benthic fauna are dominated by Chironomidae, although crayfish, isopods, and other Diptera are also found there. Stresses stemming from low dissolved oxygen and periodic drawdowns account for the low diversity and number in these domes.

The production of wood in deepwater swamps results in an abundance of substrate for invertebrates to colonize, although few studies have documented the importance of this substrate in swamps for invertebrates. Thorp et al. (1985) found that suspended *Nyssa* logs had three times as many invertebrates and twice as many taxa when they were placed in a swamp-influent stream than in the swamp itself or by its outflow stream. The swamp inflow had the highest number of mayflies (Ephemeroptera), stoneflys (Plecoptera), midges (Chironomids), and caddie flies (Trichoptera), whereas Oligochaetes were greatest in the swamp itself, supposedly because of anoxic, stagnant conditions.

Understory plants within swamps have been shown to be important to various invertebrate groups. Sampling 24 boreal forests, Hornung and Foote (2006) found that herbivorous invertebrates tended to occur in more complex vegetated areas while predatory invertebrates were more likely in areas with less plant volume. They surmised that the preference shown by predatory invertebrates was a combination of better visual conditions and lower occurrence of *Myriophyllum* spp. which contains allelopathic compounds that can decrease algae and prey abundance.

Fish

Fish are both temporary and permanent residents of alluvial river swamps. Several studies have noted the value of sloughs and backswamps for fish and shellfish spawning and feeding during the flooding season. Forested swamps often serve as a reservoir for fish when flooding ceases, although the backwaters are less than optimum for aquatic life because of fluctuating water levels and occasional low-dissolved-oxygen levels. Some fish such as bowfin (*Amia calva*), gar (*Lepisosteus* sp.), and certain top minnows (e.g., *Fundulus* spp. and *Gambusia affinis*) are better adapted to periodic anoxia through their ability to utilize atmospheric oxygen. Several species of forage minnows often dominate alluvial river swamps, where larger fish are temporary residents of the wetlands. Fish are sparse to nonexistent in the shallow cypress domes, white cedar swamps, and red maple swamps because of the lack of continuous standing water.

Reptiles and Amphibians

Reptiles and amphibians are prevalent in swamps because of their ability to adapt to fluctuating water levels. Nine or ten species of frogs are common in many southeastern cypress–gum swamps. Two of the most interesting reptiles in southeastern deepwater swamps are the American alligator and the cottonmouth moccasin. The alligator ranges from North Carolina through Louisiana, where alluvial cypress swamps and cypress strands often serve as suitable habitats. The cottonmouth, or water moccasin (*Agkistrodon piscivorus*), a venomous water snake that has a white inner mouth, is found throughout much of the range of cypress wetlands and is the topic of many a "snake story" of those who have been in these swamps. Other water snakes, particularly several species of *Nerodia*, however, are often more important in terms of number and biomass and are often mistakenly identified as cottonmouth. The snakes feed primarily on frogs, small fish, salamanders, and crayfish.

Red maple swamps are important areas in the forested northeastern United States for the breeding and feeding of reptiles and amphibians. DeGraaf and Rudis (1986) found that there were 45 species of reptiles and amphibians that required forest cover sometime during the year in New England and that of the 11 types of forests studied, red maple swamps were actually the preferred habitat of 12 of those 45 species. In a later study, DeGraaf and Rudis (1990) found that red maple swamps with streams supported twice as many individuals of reptiles and amphibians as did red maple swamps without streams, with wood frog (*Rana sylvatica*), redback salamander (*Plethodon cinereus*), and American toad (*Bufo americanus*) accounting for 90 percent of the abundance.

Ecosystem Function

Several generalizations about the ecosystem function of freshwater swamps are discussed in this section:

- Swamp productivity is closely tied to its hydrologic regime.
- Nutrient inflows, often coupled with hydrologic conditions, are major sources of influence on swamp productivity.
- Swamps can be nutrient sinks whether the nutrients are a natural source or are artificially applied.
- Decomposition of woody and non-woody material in swamps is affected by the water regime and the subsequent degree of anaerobiosis.

Primary Productivity

The importance of flood pulsing (the flood stability concept of W. E. Odum et al., 1995) to the productivity of swamps is illustrated in Figure 3-6a, where the basal-area growth of bald cypress in an alluvial river swamp in southern Illinois was strongly correlated with the annual discharge of the adjacent river. This graph suggests that higher tree productivity in this wetland occurred in years when the swamp was flooded more frequently than average or for longer durations by the nutrient-rich river. Similar correlations were also obtained when other independent variables that indicate degree of flooding were used. The importance of nutrient inflows as well as hydrologic conditions to productivity in cypress swamps in general is illustrated in Figure 3-6b. Hydrologic inflows and nutrient inflows are coupled in most swamps, so both charts in Figure 3-6 reflect the same phenomenon. There is a wide range of productivity reported for forested swamps, with almost all of the studies carried out in the southeastern United States (see Table 3-5). Primary productivity depends on hydrologic and nutrient conditions and pulsing hydrology supports more productive systems than does permanent flooding or lack of flooding.

Several studies have reported the importance of flooding to forests by linking annual tree growth with flood occurrence (Conner and Day, 1976; Mitsch and Ewel, 1979; Robertson et al., 2001; Stromberg, 2001; Anderson and Mitsch, 2008a,b). Based on the subsidy-stress model (E.P. Odum et al., 1979; E.P. Odum, 2000), floodplain tree growth should be maximized where flooding is frequent or long enough to subsidize nutrients and enhance growing conditions but not so much that floods become a physiological stress to trees. Attempts to demonstrate this model by comparing forest communities along a wetness gradient have often been inconclusive. Megonigal et al. (1997) investigated productivity in floodplain swamps throughout the southeastern United States and concluded that while permanently flooded floodplain swamps did have lower productivity, there was no evidence that sites that were seasonally pulsed were any more productive than sites that were clearly upland (see Figure 3-7). They suggest that the model developed by Mitsch and Rust (1984) (see Mitsch and Gosselink, 2007, Chapter 4) may be a more appropriate description of the productivity of forested wetlands.

In almost all of these studies, only aboveground productivity was estimated. Powell and Day (1991) made direct measurements of belowground productivity and found that it was highest in a mixed hardwood swamp ($989\,g\,m^{-2}\,yr^{-1}$) and much lower in a more frequently flooded cedar swamp ($366\,g\,m^{-2}\,yr^{-1}$), a cypress swamp ($308\,g\,m^{-2}\,yr^{-1}$), and a maple–gum swamp ($59\,g\,m^{-2}\,yr^{-1}$). These results suggest that the allocation of carbon to the root system decreases with increased flooding.

In forested wetlands that have been unaltered, annual mortality rates are often low. Conner et al. (2002) monitored annual changes in forested wetland structure between 1987 and 1999 in South Carolina and Louisiana. Tree mortality in unaltered areas was low (∼2%), but higher annual mortality (up to 16%) was observed at Louisiana sites where severe water level rise occurred. They also found that severe wind storms increased short-term mortality but these events can also lead to elevated long-term mortality rates as damaged trees eventually succumb.

a.

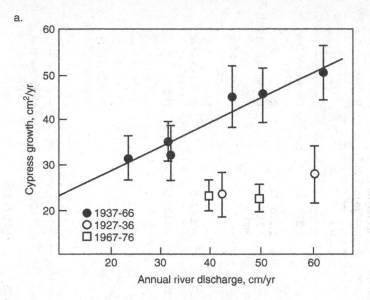

b.

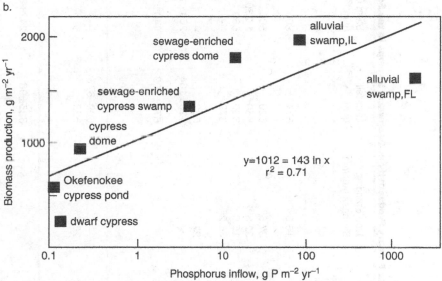

Figure 3-6   Relationships between hydrologic conditions and tree productivity in cypress swamps: *a*. increase in basal area of bald cypress trees in southern Illinois alluvial swamp as a function of river discharge for 5-year periods, and *b*. biomass production as a function of phosphorus inflow for several cypress swamps. Data points in *a*. indicate mean; bar indicates one standard error. (*a. After Mitsch et al., 1979b. After S. L. Brown, 1981*)

Table 3-5 Biomass and Net Primary Productivity of Deepwater Swamps in the Southeastern United States

Location/Forest Type	Tree Standing Biomass (kg/m²)	Litterfall (g m⁻² yr⁻¹)	Stem Growth (g m⁻² yr⁻¹)	Aboveground NPP[a] (g m⁻² yr⁻¹)	Reference
Louisiana					
Bottomland hardwood	16.5[b]	574	800	1,374	Conner and Day (1976)
Cypress–tupelo	37.5[b]	620	500	1,120	Ibid.
Impounded managed swamp	32.8[b,c]	550	1,230	1,780	Conner et al. (1981)
Impounded stagnant swamp	15.9[b,c]	330	560	890	Ibid.
Tupelo stand	36.2[b]	379	—	—	Conner and Day (1982)
Cypress stand	27.8[b]	562	—	—	Ibid.
Cypress forests (n=7)		425±164	269±117	765±245	Megonigal et al. (1997)
Bottomland hardwood forests (n=7)		670±109	440±135	1,110±177	Ibid.
North Carolina					
Tupelo swamp	—	609–677	—	—	Brinson (1977)
Floodplain swamp	26.7[d]	523	585	1,108	Mulholland (1979)
South Carolina					
Cypress forests (n=2)		385±98	242±64	625±35	Megonigal et al. (1997)
Bottomland hardwood forests (n=10)		720±98	488±122	1,208±198	Ibid.
Virginia					
Cedar swamp	22.0[b]	758	441	1,097[f]	Dabel and Day (1977), Gomez and Day (1982), Megonigal and Day (1988)
Maple gum swamp	19.6[b]	659	450	1,050[f]	Ibid.

Type					Reference
Cypress swamp	34.5[b]	678	557	1,176[j]	Ibid.
Mixed hardwood swamp	19.5[b]	652	249	831[j]	Ibid.
Georgia					
Nutrient-poor cypress swamp	30.7[e]	328	353	681	Schlesinger (1978)
Illinois					
Floodplain forest	29.0	—	—	1,250	F.L. Johnson and Bell (1976)
Floodplain forest		491	177	668	S.L. Brown and Peterson (1983)
Cypress–tupelo swamp	45[d]	348	330	678	Mitsch (1979), Dorge et al. (1984)
Ohio					
Bottomland hardwood forest(before and after hydrologic restoration)				531-641 (pre-restoration) 807 ± 86 (post-restoration)	Cochran (2001); Anderson and Mitsch (2008a,b)
Kentucky					
Cypress–ash slough	31.2	136	498	634	Mitsch et al. (1991)
Cypress swamp	10.2	253	271	524	Ibid.
Stagnant cypress swamp	9.4	63	142	205	Ibid.
Bottomland forest	30.3	420	914	1,334	Ibid.
Bottomland forest	18.4	468	812	1,280	Ibid.
Florida					
Cypress–tupelo (6 sites)	19 ± 4.7[f]	—	289 ± 58[f]	760[g]	Mitsch and Ewel (1979)
Cypress–hardwood (4 sites)	15.4 ± 2.9[f]	—	336 ± 76[f]	950[g]	Ibid.
Pure cypress stand (4 sites)	9.5 ± 2.6[f]	—	154 ± 55[f]	—	Ibid.
Cypress–pine (7 sites)	10.1 ± 2.1[f]	—	117 ± 27[f]	—	Ibid.

(continues)

Table 3-5 (continued)

Location/Forest Type	Tree Standing Biomass (kg/m²)	Litterfall (g m⁻² yr⁻¹)	Stem Growth (g m⁻² yr⁻¹)	Aboveground NPP[a] (g m⁻² yr⁻¹)	Reference
Floodplain swamp	32.5	521	1,086	1,607	S. L. Brown (1978)
Natural dome[h]	21.2	518	451	969	Ibid.
Sewage dome[i]	13.3	546	716	1,262	Ibid.
Scrub cypress	7.4	250	—	—	S. L. Brown and Lugo (1982)
Drained strand	8.9	120	267	387	Carter et al. (1973)
Undrained strand	17.1	485	373	858	Ibid.
Larger cypress strand	60.8	700	196	896	Duever et al. (1984)
Small tree cypress strand	24.0	724	818	1,542	Ibid.
Sewage enriched cypress strand	28.6	650	640	1,290	Nessel (1978)

[a]NPP = net primary productivity = litterfall + stem growth.
[b]Trees defined as >2.54 cm DBH (diameter at breast height).
[c]Cypress, tupelo, ash only.
[d]Trees defined as >10 cm DBH.
[e]Trees defined as >4 cm DBH.
[f] Average ± std error for cypress only.
[g]Estimated.
[h]Average of five natural domes.
[i]Average of three domes; domes were receiving high nutrient wastewater.
[j]Aboveground NPP is less than sum of litterfall plus stem growth because some stem growth is measured as litterfall.

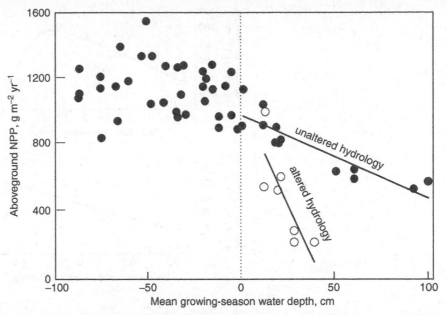

Figure 3-7 The relationship between net primary productivity of floodplain forests and mean growing season water depth in bottomland hardwood forests of the southeastern United States. (*After Megonigal et al., 1997*)

Pollock et al. (1998) showed that species richness in a riparian wetland could be influenced by productivity and disturbance frequency. Examining 16 sites (representing a wide range of wetland types) in the Kadashan River basin in Alaska, they found that species-rich sites had low to intermediate productivity and intermediate disturbance (flooding) frequencies (see Figure 3-8). Low species richness sites tended to have low or high flood frequencies with low productivity. Spatial variation of flood disturbance frequencies also promoted greater diversity. Their findings supported Huston's (1994) dynamic equilibrium model of species diversity.

Energy Flow

The energy flow of deepwater swamps is dominated by primary productivity of the canopy trees. Energy consumption is primarily accomplished by detrital decomposition. Significant differences exist, however, between the energy flow patterns in low-nutrient swamps such as dwarf cypress swamps and cypress domes and high-nutrient swamps such as alluvial cypress swamps (see Table 3-6). All of the cypress wetlands are autotrophic—productivity exceeds respiration. Gross primary productivity, net primary productivity, and net ecosystem productivity are highest in the alluvial river swamp that receives high-nutrient inflows. Buildup and/or export of organic matter are characteristic of all of these deepwater swamps but are most characteristic of the alluvial swamp. There are few allochthonous inputs of energy to

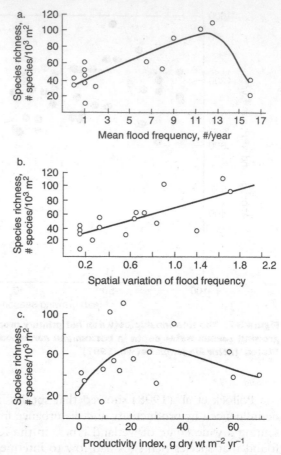

Figure 3-8 Relationship between plant species richness and a. flooding frequency; b. spatial variation of flooding frequency; and c. productivity for different riparian types in the Kadashan River basin in southeastern Alaska. Spatial variation of flood frequency was calculated as the coefficient of variation of the mean flood frequency based on 55 microplot elevations measured at each site. (*After Pollock et al., 1998*)

the low-nutrient wetlands, and energy flow at the primary producer level is relatively low. The alluvial cypress–tupelo swamp depends more on allochthonous inputs of nutrients and energy, particularly from runoff and river flooding. In alluvial deepwater swamps, productivity of aquatic plants is often high, whereas aquatic productivity in cypress domes is usually low.

Nutrient Budgets

The functioning of forested wetlands as nutrient sinks was first suggested by Kitchens et al. (1975) in a preliminary winter–spring survey of a swamp forest alluvial river swamp complex in South Carolina. They found a significant reduction in phosphorus

Table 3-6 Estimated Energy Flow (kcal m^{-2} day^{-1}) in Selected Florida Cypress Swamps[a]

Parameter	Dwarf Cypress Swamp	Cypress Dome	Alluvial River Swamp
Gross primary productivity[b]	27	115	233
Plant respiration[c]	18	98	205
Net primary productivity	9	17	28
Soil or water respiration	7	13	18
Net ecosystem productivity	2	4	10

[a]Assume 1 g C = 10 kcal.
[b]Assumes GPP = net daytime photosynthesis + nighttime leaf respiration.
[c]Plant respiration = 2 × (nighttime leaf respiration) + stem respiration + knee respiration.
Source: Adapted from S. L. Brown (1981).

as the waters passed over the swamp and assumed this to be the result of biological uptake by aquatic plant communities. In a similar study in Louisiana, J. W. Day et al. (1977) found that nitrogen was reduced by 48 percent and phosphorus decreased by 45 percent as water passed through a lake–swamp complex of Barataria Bay to the lower estuary. They attributed this decrease in nutrients to sediment interactions, including nitrate storage/denitrification and phosphorus adsorption to the clay sediments. Since then, countless studies have illustrated the potential of forested wetlands for nutrient removal, including several studies in Louisiana (Mitsch and Day, 2004; Day et al., 2004).

Nutrient budgets of deepwater swamps vary from "open" alluvial river swamps that receive and export large quantities of materials to "closed" cypress domes that are mostly isolated from their surroundings (see Table 3-7). Mitsch et al. (1979) developed a nutrient budget for an alluvial river swamp in southern Illinois and

Table 3-7 Phosphorus Inputs to Deepwater Swamps (g P m^{-2} yr^{-1})

Swamp	Rainfall	Surface Inflow	Sediments from River Flooding	Reference
Florida				
Dwarf cypress	0.11	–[a]	0	S. L. Brown (1981)
Cypress dome	0.09	0.12	0	Ibid.
Alluvial river swamp	–[a]	–[a]	3.1	Ibid.
Southern Illinois				
Alluvial river swamp	0.11	0.1	3.6	Mitsch et al. (1979)
North Carolina				
Alluvial tupelo swamp	0.02–0.04	0.01–1.2	0.2	Yarbro (1983)

[a]Not measured.

found that 10 times more phosphorus was deposited with sediments during river flooding ($3.6\,g\ P\ m^{-2}\ yr^{-1}$) than was returned from the swamp to the river during the rest of the year. The swamp was a sink for a significant amount of phosphorus and sediments during that particular year of flooding, although the percentage of retention was low (3–4.5 percent) because a very large volume of water passed over the swamp during flooding conditions. Noe and Hupp (2005) evaluated net nutrient accumulation in floodplain forests along rivers contributing to the Chesapeake Bay. Mean accumulation rates for C ranged from 61 to $212\,g\ m^{-2}\ yr^{-1}$, N ranged from 3.5 to $13.4\,g\ m^{-2}\ yr^{-1}$, and P ranged from 0.2 to $4.1\,g\ m^{-2}\ yr^{-1}$. Watershed land use was a significant factor in their study with the greatest accumulation of sediment and nutrients occurring along the Chickahominy River, downstream from the urban metropolitan area of Richmond, Virginia.

Freshwater Marshes

This category of wetlands includes a diverse group of wetlands characterized by (1) emergent soft-stemmed aquatic plants such as cattails, arrowheads, pickerelweed, reeds, and several other species of grasses and sedges; (2) a shallow-water regime; and (3) generally shallow to nonexistent peat deposits. We estimate that there are about 95 million ha of freshwater marshes in the world (refer to Table 3-1). Therefore, inland marshes represent less than 20% of the total amount of wetlands in the world. The Okavango Delta in Botswana, the Danube and Volga Deltas in Eastern Europe, the Mesopotamian Marshlands in Iraq, the Everglades in Florida, the Prairie Pothole region of the United States and Canada, and the Pantanal in South America are all examples of regions with extensive expanses of freshwater marshes. Freshwater marshes are estimated to cover about 9.6 million ha in the coterminous United States (see Table 3-1).

The freshwater marshes discussed in this section of Chapter 3 are a diverse group. Nevertheless, they can be treated as a unit because they are nontidal, freshwater systems dominated by grasses, sedges, and other freshwater emergent hydrophytes. Otherwise, they differ in their geological origins and in their driving hydrologic forces, and they vary in size from the small pothole marshes of less than a hectare in size to the immense sawgrass monocultures of the Florida Everglades.

There are few accurate measures of how many freshwater marshes there are in the world for several reasons. First, they are often ephemeral or convert to other types of wetlands such as unvegetated flats or forested wetlands over a relatively short period of time. Second, they can be confused with and miscounted with peatlands, particularly fens. Third, freshwater marshes have such a wide number of possible dominant vegetation covers and water depths (from saturated soils to 1 m of water depth) that their classification and inventory is very difficult.

Hydrology

As with any other wetland, the flooding regime, or hydroperiod, of freshwater marshes determines their ecological character. The critical factors that determine the character of these wetlands are the presence of excess water and sources of water other than direct precipitation. Excess water occurs either when precipitation exceeds evapotranspiration or when the watershed draining into the marsh is large enough to provide adequate runoff or groundwater inflow. The hydroperiods of several freshwater marsh systems are illustrated in Mitsch and Gosselink (2007) in Chapter 4. Along seacoasts, water levels tend to be stable over the long term because of the influence of the ocean. Water levels in inland marshes, in contrast, are much more controlled by the balance between precipitation and evapotranspiration, especially for marshes in small watersheds that are affected by restricted throughflow. Water levels of marshes, such as those found along the Laurentian Great Lakes, are generally stable but are influenced by the year-to-year variability of lake levels and whether the wetland is diked or open to the lake. Many marshes such as wet meadows, sedge meadows, vernal pools, and even prairie potholes dry down seasonally, but the plant species found there reflect the hydric conditions that exist during most of the year. The seasonality of these marshes occurs because they are primarily fed by runoff and precipitation.

Some marshes intercept groundwater supplies. Their water levels, therefore, reflect the local water table, and the hydroperiod is less erratic and seasonal. These types of marshes, such as those found in the prairie pothole region of North America, can be either recharge or discharge wetlands. Other marshes collect surface water and entrained nutrients from watersheds that are large enough to maintain hydric conditions most of the time. For example, overflowing lakes supply water and nutrients to adjacent marshes, and riverine marshes are supplied by the rise and fall of the adjacent river. Because river and stream discharge, lake levels, and precipitation are often notoriously variable due to weather shifts from year to year, the water regime of most inland marshes also varies in a way that is predictable only in a statistical sense. Those marshes that are fed by groundwater in addition to one or more of the previous sources are less variable, but even groundwater levels vary somewhat with season and often dramatically with climate shifts or human use.

Even in the same region, water levels can respond differently to shifts in the balance between precipitation and evapotranspiration (see Figure 3-9). Terms such as ephemeral, temporary, seasonally semi-permanent, and permanent can be used to describe freshwater marshes. In addition, marshes can move through several of these classes over the span of a few years. Thus, a marsh that would ordinarily be considered permanent might be in a "drawdown" phase that gives the appearance of an ephemeral marsh.

Biogeochemistry

The water and soil chemistry of freshwater marshes is dominated by a combination of mineral soils rather peat soils, overlain with autochthonous inputs of organic matter from the productivity of the vegetation. Given these conditions, there is still a wide range of chemical possibilities for the water and soil in freshwater marshes (see Figure 3-10). Conductivity as a measure of general salinity, can range from less than 100 μS cm^{-1} in soft-water freshwater marshes dominated by rainfall to over

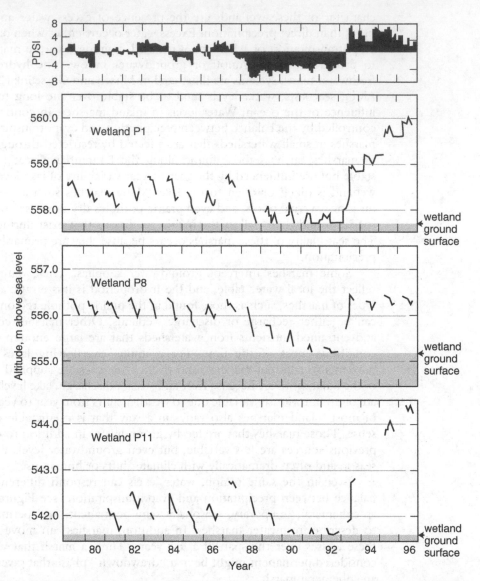

Figure 3-9 Water-level patterns in three wetlands in the prairie pothole region of North America, illustrating the uneven effect that climate has on supposedly similar wetlands in the same region, probably due to nonuniform groundwater effects on the wetlands. PDSI is the Palmer Drought Severity Index and is a relative measure of climatic "wetness." Its value decreases during drought conditions. (*After LaBaugh et al., 1996*)

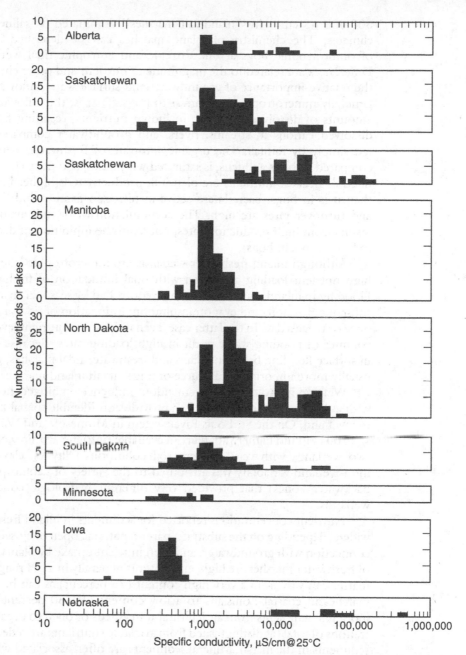

Figure 3-10 Frequency distribution of salinity, as measured by specific conductance, for prairie lakes and wetlands in the United States and Canada. Concentrations shown for Nebraska are typical of inland salt marshes. (*After LaBaugh, 1989*)

300,000 μS cm^{-1} in "inland salt marshes" dominated by saline seeps in semi-arid climates. The chemistry of inland marshes can be described in contrast to that of ombrotrophic bogs at one extreme and eutrophic tidal wetlands at the other. Differences are related to the magnitude of nutrient and other chemical inputs and to the relative importance of groundwater and surface water inflow. Inland marshes are generally minerotrophic in contrast to bogs. That is, the inflowing water has higher amounts of dissolved materials, including nutrients, resulting from the presence of dissolved cations in streams, rivers, and groundwater compared to bogs that are fed simply by rainfall. The organic substrate of freshwater marshes, while shallow compared to bogs and fens, is saturated with bases, and the pH, as a result, is close to neutral. Because nutrients are plentiful, productivity is higher in freshwater marshes than it is in bogs, bacteria are active in nitrogen fixation and litter decomposition, and turnover rates are high. The accumulation of organic matter that does occur results from high production rates, not from the inhibition of decomposition by low pH (as occurs in bogs).

Although inland freshwater wetlands are minerotrophic, they generally lack the high nutrient loading associated with tidal inundation of freshwater tidal marshes. Flooding in inland marshes tends to be controlled by seasonal changes in local rainfall, subsequent runoff, and evapotranspiration; inundation of tidal areas occurs regularly, once or twice a day. In the latter case, even if tidewater nutrient levels are low, the large volumes of flooding water result in high loading rates. Because of these differences in surface flooding between inland and freshwater tidal marshes, groundwater flow is usually more important as a source of nutrients in inland marshes.

Wetland hydrology has a tremendous influence on biogeochemistry and, coupled with a diverse geomorphology, can provide considerable spatial heterogeneity within the wetland. On the St. Louis River system in Minnesota and Wisconsin, Johnston et al. (2001) found higher variability in available nutrients within wetlands than between two wetlands with very different soil conditions (silty vs. clayey). The significant intra-wetland variability was attributed to the variety of geomorphic features (levees, backwater zones) that provided very different hydrologic conditions within both wetlands.

Nutrient concentrations reported for sediments in inland freshwater marshes vary widely, depending on the substrate, parent material, open or closed nature of the basin, connection with groundwater, and even nutrient uptake by plants. Ion concentrations of freshwater marshes are high, and water is generally in a pH range of 6 to 9. Organic matter can vary from a very high content (75 percent) as can be found in freshwater marshes in coastal Louisiana to a low content (10–30 percent) in marshes fed by inorganic sediments from agricultural watersheds or open to organic export. Concentrations of total (as distinguished from available) nutrients are reflections of the kinds of sediments in the marsh. Mineral sediments are often associated with high phosphorus content, for example, whereas total nitrogen is closely correlated to organic content. Dissolved inorganic nitrogen and phosphorus—the elements that most often limit plant growth—often vary seasonally from very low concentrations in the summer, when plants take them up as rapidly as they become available, to high concentrations in the winter, when plants are dormant but mineralization continues in the soil.

The anaerobic conditions in marshes and other wetlands give them the potential to emit considerable amounts of nitrous oxide (N_2O) and methane (CH_4), both of which are considered important greenhouse gases. Temperature and diffusion rates through water influence the net emission of these gases and therefore shallow wetlands will often have greater emissions than open water bodies. In a boreal lake in Finland, Huttunen et al. (2003) estimated that the littoral zone, which consisted of 26% of the total lake surface area, was responsible for most of the N_2O emissions from the lake.

Methane emission from wetlands has been studied extensively because it has a much greater atmospheric warming potential than N_2O or CO_2. Brix et al. (2001) examined whether *Phragmites* marshes in Europe could be considered a net source or sink for greenhouse gases given that marshes assimilate CO_2 and store carbon while emitting CH_4. They found that when these marshes are evaluated over a shorter time period (<60 years) these wetlands could be considered a net source of greenhouse gases based on their emission of CH_4 and CO_2 relative to C fixation. However, CH_4 does not persist in the atmosphere indefinitely and if these marshes are evaluated over a longer time period (>100 years) the balance shifts and these marshes can be considered a net sink for greenhouse gases.

In many parts of the world where arid climates persist, inland marshes can be saline rather than fresh water. These marshes, then, have characteristics of both coastal salt marshes (because of the salinity) and inland marshes (because they are not tidal). Good examples of these kinds of marshes are the fringe marshes around the Great Salt Lake in Utah, the Salton Sea in California, and the Caspian Sea in Eastern Europe. There are also inland "salt marshes" in places such as eastern Nebraska where these marshes, some as large as 15 to 20 ha, are still found northwest of Lincoln along Salt Creek and its tributaries. These marshes are the result of saline seeps from deep groundwater and the consistently high evapotranspiration/precipitation ratio of the region.

Vegetation

The vegetation of fresh inland marshes has been detailed in many studies. The dominant species vary from place to place, but the number of genera common to all locations in the temperate zone is quite remarkable. Common species include the graminoids *Phragmites australis* (= *P. communis*; reed grass), *Typha* spp. (cattail), *Sparganium eurycarpum* (bur reed), *Zizania aquatica* (= *Z. palustris*; wild rice), *Panicum hemitomon, Cladium jamaicense*; sedges *Carex* spp., *Schoenoplectus tabernaemontani* (= *Scirpus validus*; bulrush), *Scirpus fluviatilis* (river bulrush); and *Eleocharis* spp. (spike rush). In addition, broad-leaved monocotyledons such as *Pontederia cordata* (pickerelweed) and *Sagittaria* spp. (arrowhead) are frequently found in freshwater marshes. Herbaceous dicotyledons are represented by a number of species, typical examples of which are *Ambrosia* spp. (ragweed) and *Polygonum* spp. (smartweed). Frequently represented also are such ferns as *Osmunda regalis* (royal fern) and *Thelypteris palustris* (marsh fern), and the horsetail, *Equisetum* spp. One of the most productive species in the world is the tropical sedge *Cyperus papyrus*, which flourishes in marshes and on floating mats in southern and eastern Africa.

Marsh Vegetation Zonation

These typical plant species do not occur randomly mixed together in marshes. Each has its preferred habitat. Different species often occur in rough zones on slight gradients, especially flooding gradients. Figure 3-11a illustrates the typical distribution of species along an elevation gradient in a midwestern North American freshwater marsh. Sedges (e.g., *Carex* spp., *Scirpus* spp.), rushes (*Juncus* spp.), and arrowheads (*Sagittaria* spp.) typically occupy the shallowly flooded edge of a pothole. Two species of cattail (*Typha latifolia* and *T. angustifolia*) are common. The narrow-leaved species (*T. angustifolia*) is more flood tolerant than the broad-leaved cattail (*T. latifolia*) and may grow in water up to 1 m deep. The deepest zone of emergent plants is typically vegetated with hardstem bulrush (*Scirpus acutus*) and softstem bulrush (*Schoenoplectus tabernaemontani*). Beyond these emergents, floating-leaved and submersed vegetation will grow, the latter to depths dictated by light penetration. Typical floating-leaved aquatic hydrophytes include rhizomatous plants such as water lilies (*Nymphaea tuberosa* or *N. odorata*), water lotus (*Nelumbo lutea*), and spatterdock (*Nuphar advena*), and stoloniferous plants such as water shield (*Brasenia schreberi*) and smartweed (*Polygonum* spp.). Submersed hydrophytes include coontail (*Ceratophyllum demersum*), water millfoil (*Myriophyllum* spp.), pondweed (*Potamogeton* spp.), wild celery (*Vallisneria americana*), naiad (*Najas* spp.), bladderwort (*Utricularia* spp.), and waterweed (*Elodea canadensis*).

A unique structural feature of prairie pothole marshes is the 5- to 20-year cycle of dry marsh, regenerating marsh, degenerating marsh, and lake that is related to periodic droughts. During drought years, standing water disappears. Buried seeds in the exposed mud flats germinate to grow a cover of annuals (*Bidens, Polygonum, Cyperus, Rumex*) and perennials (*Typha, Scirpus, Sparganium, Sagittaria*). When rainfall returns to normal, the mud flats are inundated. Annuals disappear, leaving only the perennial emergent species. Submersed species (*Potamogeton, Najas, Ceratophyllum, Myriophyllum, Chara*) also reappear. For the next year or more, during the regenerating stage, the emergent population increases in vigor and density. After a few years, however, these populations begin to decline. The reasons are poorly understood, but often muskrat populations explode in response to the vigorous vegetation growth. Their nest and trail building can decimate a marsh. Whatever the reason, in the final stage of the cycle, there is little emergent marsh; most of the area reverts to an open shallow lake or pond, setting the stage for the next drought cycle. Wildlife use of these wetlands follows the same cycle. The most intense use occurs when there is good interspersion of small ponds with submersed vegetation and emergent marshes with stands diverse in height, density, and potential food.

Figure 3-11b shows the plant zonation of a freshwater marsh/littoral zone in sub-Saharan Africa. The shallow flooded emergent zone is dominated by *Typha, Phragmites*, and *Cyperus papyrus*. *Typha* taxonomy in Africa has been somewhat confused, but it now appears that there are two distinct species—*T. domingensis* Pers. *sensu lato* in tropical and warm-temperate climates and *T. capensis* Rohrb. in more temperate climates of northern and southern Africa (Denny, 1993). *Cyperus papyrus* grows in the emergent zone but also develops floating islands when it breaks

a.

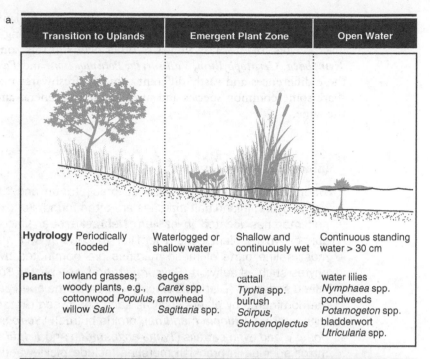

	Transition to Uplands	Emergent Plant Zone		Open Water
Hydrology	Periodically flooded	Waterlogged or shallow water	Shallow and continuously wet	Continuous standing water > 30 cm
Plants	lowland grasses; woody plants, e.g., cottonwood *Populus* willow *Salix*	sedges *Carex* spp. arrowhead *Sagittaria* spp.	cattail *Typha* spp. bulrush *Scirpus, Schoenoplectus*	water lilies *Nymphaea* spp. pondweeds *Potamogeton* spp. bladderwort *Utricularia* spp.

b.

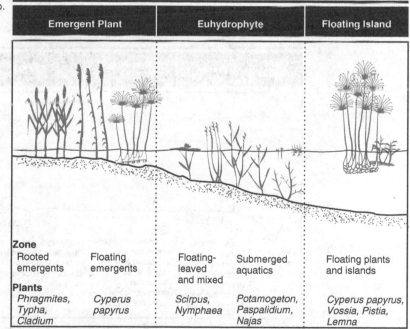

	Emergent Plant		Euhydrophyte		Floating Island
Zone	Rooted emergents	Floating emergents	Floating-leaved and mixed	Submerged aquatics	Floating plants and islands
Plants	*Phragmites, Typha, Cladium*	*Cyperus papyrus*	*Scirpus, Nymphaea*	*Potamogeton, Paspalidium, Najas*	*Cyperus papyrus, Vossia, Pistia, Lemna*

Figure 3-11 Cross-sections of vegetation through freshwater marshes, indicating plant zones according to water depth and typical plants found in each zone for a. temperate-zone midwestern North America and b. sub-Saharan Africa. Note floating marsh islands in African wetlands. (b. *After Denny, 1993*)

free from the shoreline. The euhydrophyte (true water plant) zone in Figure 3-11b was coined by Denny (1985) to refer to rooted, floating-leaved, and submerged macrophytes. In Africa, examples of other plants in this zone are *Chara, Fontinalis, Nymphaea, Ceratophyllum, Vallisneria, Potamogeton*, and *Paspalidium*. Thus, despite these differences and vastly different climates, freshwater marshes around the world share some common species and many common genera, and are functionally much the same.

Inland Salt Marshes

Where evapotranspiration exceeds precipitation and/or saline groundwater seeps occur, inland salt marshes are often found. For example, the Nebraska salt marshes located in eastern Nebraska near Lincoln (see Figure 3-12) support many plant genera familiar to coastal salt marsh ecologists. The most saline parts of these marshes are dominated by salt-tolerant macrophytes such as saltwort (*Salicornia rubra*), sea blight (*Suaeda depressa*), and inland saltgrass (*Distichlis spicata*), whereas the open ponds and their fringes are dominated by plants such as sago pondweed (*Potamogeton pectinatus*), wigeon grass (*Ruppia maritima*), prairie bulrush (*Scirpus maritimus* var. *paludosus*), and even cattails (*Typha angustifolia* and *T. latifolia*). In California, the major species in brackish marshes include pickleweed (*Salicornia virginica*) and alkali bulrush (*Scirpus robustus*) (deSzalay and Resh, 1996, 1997).

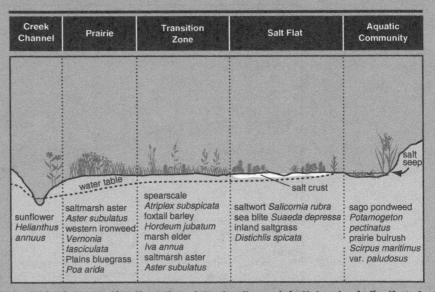

Figure 3-12 **Cross-section through an inland salt marsh in Nebraska, indicating plant zones and typical plants found in each zone. (*After Farrar and Gersib, 1991*)**

Reed Die-Back

A vegetation dynamic that has been affecting marshes in Europe and elsewhere is the phenomenon known as *reed dieback*. Since the early 1950s, scientists have noted a retreat of the common reed *Phragmites australis* around lakes throughout Europe. Several factors that have been suggested to cause the loss of reeds include:

1. Direct destruction through human development
2. Mechanical damage due to waves, winds, boat traffic, etc.
3. Grazing by animals such as geese, swans, coots, muskrats, and cattle
4. Lake eutrophication and sedimentation
5. Lake level manipulation and bank erosion
6. Replacement of reeds by species such as *Typha* sp. and *Glyceria maxima* (Ostendorp et al., 1995)

In what may seem surprising to Eastern United States coastal marsh managers, who have been trying to reduce the advance of *Phragmites* into tidal freshwater and brackish marshes for decades, the loss of *Phragmites* beds around lakes in Europe is considered to be a major environmental concern. Their loss has been cited as a main reason for the general deterioration of the lake littoral zone, including increased erosion and deposition of shorelines; loss of bird species such as warblers, bitterns, grebes, and herons; uprooting of shoreline trees and bushes; loss of submersed vegetation; and reduction in fish populations.

Seed Banks and Germination

Seed banks and fluctuating water levels interact in complicated ways to produce vegetation communities in freshwater marshes (Mitsch and Gosselink, 2007). As a general rule, seed germination is maximized under shallow water or damp soil conditions, after which many perennials can reproduce vegetatively into deeper water. For example, fluctuating water levels along the Great Lakes allowed greater diversity of plant types and species in the coastal marshes and these marshes sometimes have a density of buried seeds an order of magnitude greater than that of the more studied prairie marshes (Keddy and Reznicek, 1986).

The importance of the timing of flooding and drying was illustrated in an experiment in Ohio involving uniform seed banks subjected to several hydroperiods. Although plant density and aboveground biomass were not affected by the different hydroperiods, species composition, diversity, and richness were affected (see Figure 3-13). Highest richness and diversity occurred in continuously moist soils. Flooding followed by a drawdown to moist soils, as is a typical hydroperiod in the midwestern United States, encouraged obligate wetland species, whereas moist soil conditions followed by flooding encouraged the growth of fewer wetland species and more annual species.

The importance of other variables, particularly nutrients, for the germination and success of freshwater marsh plants is not as well understood. In another series

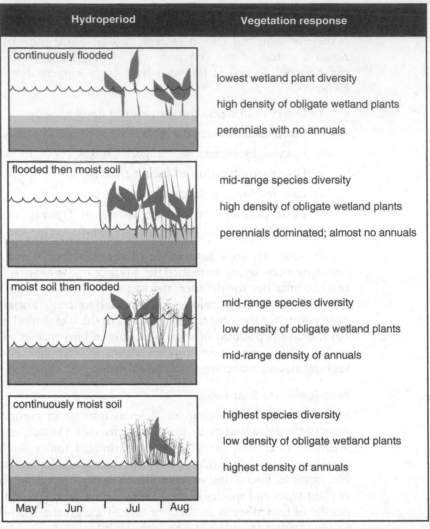

Hydroperiod	Vegetation response
continuously flooded	lowest wetland plant diversity high density of obligate wetland plants perennials with no annuals
flooded then moist soil	mid-range species diversity high density of obligate wetland plants perennials dominated; almost no annuals
moist soil then flooded	mid-range species diversity low density of obligate wetland plants mid-range density of annuals
continuously moist soil	highest species diversity low density of obligate wetland plants highest density of annuals

May   Jun   Jul   Aug

Figure 3-13 Experimental results of four different growing season hydroperiods on a common seed bank of annual and perennial freshwater marsh plants in Ohio. Continuously deep water favored low density of obligate wetland perennials; flooded then moist, typical of natural hydroperiod for the midwestern United States, favored perennial-dominated and diverse wetland plants; moist then flooded, typical of some managed marshes around the Laurentian Great Lakes, favored fewer obligate wetland plants and higher density of annuals; continuously flooded soil had highest diversity but fewest obligate wetland plants and highest density of annuals.

of experiments, Gerritsen and Greening (1989) illustrated the importance of water levels and nutrient conditions for seed germination in seed banks from two marshes in the Okefenokee Swamp in Georgia. Although water level was the important variable in determining which plants germinated, nitrogen was shown to limit marsh plant growth in drawdown conditions, and phosphorus limited the growth of a few species while they were inundated. In a series of experiments using natural (many species) and synthetic seed banks (two species only) under different hydrologic and nutrient conditions, Willis and Mitsch (1995) found similar effects of water level on germination but were unable to show the importance of nutrient additions.

Vegetation Diversity

The particular species found in freshwater wetlands are also determined by many other environmental factors. Nutrient availability determines, to a large degree, whether a wetland site will support mosses or angiosperms (i.e., whether it is a bog or a marsh) and what the species diversity will be. It is not obvious for freshwater marshes that highly fertile wetlands are highly diverse. In fact, many studies of freshwater marsh plant diversity published in the literature (e.g., Wheeler and Giller, 1982; Vermeer and Berendse, 1983; R. T. Day et al., 1988; D. R. J. Moore and Keddy, 1989; Moore et al., 1989; Mitsch et al., 2005b) suggest the opposite conclusion. Moore et al. (1989) contrasted several fertile and infertile sites (as measured by the plant standing crop) in eastern Ontario and found the greatest species richness in marshes that had peak biomass between 60 and 400 g/m^2 and much less plant richness at higher plant standing crop (>600 g/m^2) (see Figure 3-14). They also found rare species only at

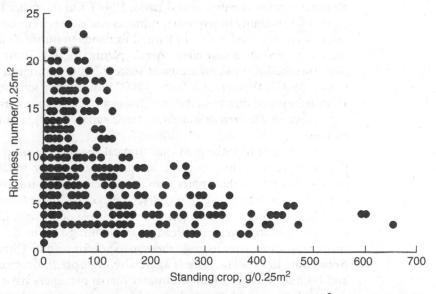

Figure 3-14 Species richness versus vegetation biomass in 0.25-m^2 quadrats from three wetland areas in Ontario, Quebec, and Nova Scotia. (*After Moore et al., 1989*)

the infertile sites, suggesting that the conservation of infertile wetlands should be part of overall wetland management strategies.

Plant species also change with latitude—that is, as temperatures increase or decrease and the winters become more or less severe. Although the same genera may be found in the Tropics and in the Arctic, the species are usually different, reflecting different adaptations to cold or heat. Because of long isolation and evolution, the flora of the North American continent differs at the species level from the European flora. Finally, soil salt, even in low concentration, determines the species found on a site. Because many inland marshes are potholes that collect water that leaves only by evaporation, salts may become concentrated during periods of low precipitation, adversely affecting the growth of salt-intolerant species. In a review of 90 emergent wetlands along the U.S. Great Lakes, Johnston et al. (2007) revealed that plant forms were often indicators of wetland soil types. They found submerged aquatic vegetation tended to indicate silty soils, free-floating plants indicated clay soils, and graminoids indicated sandy soils.

Invasive Species

Nonnative plant species are often a part of the vegetation of freshwater marshes, particularly in areas that have been disturbed. Plants such as *Eichhornia crassipes* (water hyacinth), *Salvinia molesta* (salvinia), and *Alternanthera philoxeroides* (alligator weed) have invaded tropical and subtropical regions of the world. In the southeastern United States, *E. crassipes* is considered to be a nuisance plant because of its prolific growth rate. The free-floating aquatic plant can double the area that it covers in two weeks and has choked many waterways that have received high nutrient loads for almost a century (Penfound and Earle, 1948). On the other hand, the plant has been praised for its ability to sequester nutrients and other chemicals from the water and has often been proposed as part of natural wetlands to purify wastewater. Although there are many theories about alien aquatic plants, there is some validity to the concept that disturbed ecosystems are most susceptible to biological invasions (Mitchell and Gopal, 1991). Werner and Zedler (2002) found that sediment accumulation within Wisconsin sedge meadows reduced tussock microtopography, promoted invasion by *Typha* spp. or *Phalaris arundinacea* (reed canary grass), and reduced wetland species richness.

It has been hypothesized that tropical regions are more susceptible to invasion than temperate regions because invading plants grow much more rapidly and are more noticeable in the tropics than in temperate latitudes. In the freshwater marshes of the St. Lawrence and Hudson River valleys and in the Great Lakes region of North America, however, *Lythrum salicaria* (purple loosestrife), a tall purple-flowered emergent hydrophyte, has spread at an alarming rate in this century, causing much concern to those who manage these marshes for wildlife (Stuckey, 1980; Balogh and Bookhout, 1989). The plant is aggressive in displacing native grasses, sedges, rushes, and even *Typha* spp. Many freshwater marsh managers have implemented programs designed to control purple loosestrife by chemical and mechanical means. Other aquatic aliens such as the submersed *Hydrilla verticillata*, a plant native to Africa,

Asia, and Australia, and *Myriophyllum spicatum* have invaded open, shallow-water marshes in the United States (Steward, 1990; Galatowitsch et al., 1999) but rarely compete well with emergent vegetation.

Phragmites australis (Cav.) Trin. ex. Steud., common reed, is now considered an invasive species in eastern North America, particularly along the Atlantic coastline and around the Laurentian Great Lakes, even though the plant has been in North America for more than 3,000 years, because of its aggressive expansion through brackish (salinity < 5 ppt) and freshwater marshes, especially in the past 50 years. Its spread is attributed to increased disturbances, spread of more aggressive varieties from other parts of the world, including Europe, and change in hydrology and salinity patterns in coastal estuarine systems (Philipp and Field, 2005). Major resources are used to control this plant from spreading in eastern North America with techniques such as burning and herbicide application. There is irony in the fact that the plant is taking over North American wetlands while reed-dieback of the same species in Europe (see above) is the main concern for this species there. More recently, two subspecies of this species have been identified in North America (Saltonstall et al., 2004). *Phragmites australis* subsp. *americanus* was identified as a race native to North America and *Phragmites australis* subsp. *australis* is the aggressive non-native race now abundant in North America but generally considered to be of European origin.

Consumers

Perhaps one reason that small marshes of the prairie region and the western high plains harbor such a rich diversity of organisms and wildlife is that they are often natural islands in a sea of farmland. Cultivated land does not provide a diversity of either food or shelter, and many animals must retreat to the marshes, which have become their only natural habitats. In cases where flow from watersheds is seasonal, freshwater marshes can serve as biological and hydrologic "oases" during low-flow and drought conditions.

Invertebrates

Invertebrates, similar to amphibians, are the links between plants and their detritus, on the one hand, and animals such as fish, ducks and other birds, and even several mammals, on the other. Insects make up much of the invertebrate taxa in freshwater marshes and their composition is often dictated by wetland hydrology and vegetation. In their literature review, Batzer and Wissinger (1996) noted that temporary pools tend to be diverse with beetle and midge communities. Insect communities are often productive because of the alternating wet-dry conditions and many communities are regulated by biotic interactions. As marshes become more perennial, vegetation for habitat structure and as a decaying substrate becomes more important for invertebrates. In open water sections, certain benthos and nektonic insects may also be important.

The most conspicuous invertebrates are the true flies (Diptera), which often make one's life miserable in the marsh. These include midges, mosquitoes, and crane flies.

However, in the larval stage, many of the insects are benthic. Midge larvae, which are called *bloodworms* because of their rich red color, "are found submerged in bottom soils and organic debris, serving as food for fish, frogs, and diving birds. When pupae surface and emerge as adults, they are exploited as well by surface-feeding birds and fish" (Weller, 1994). Odonata, represented by dragonflies and damselflies, are a notable feature of freshwater marshes; their very presence generally indicates good water quality. Crustaceans such as crayfish and mollusks such as snails can be common in some freshwater marshes. The former are food for large fish and mammals alike, whereas the latter are often found grazing on mats of filamentous algae.

Temporal cycles and spatial patterns of invertebrate species and concentrations reflect the natural seasonal cycle of insect growth and emergence superimposed on the vegetation cycles. McLaughlin and Harris (1990) investigated insect emergence from diked and undiked marshes along Lake Michigan and found more insects, more insect biomass, and a greater number of taxa in diked marshes and the greatest numbers and biomass in the sparsely vegetated zones of the wetlands rather than in open water or dense vegetation. Kulesza and Holomuzki (2006) examined growth and survival of the detritivorous Amphipod *Hyalella azteca* from a Lake Erie marsh. They compared its use of *Typha angustifolia* and *Phragmites australis* as substrate and found that both plants supported adequate fungi growth and the amphipods performed equally well.

Mammals

A number of mammals inhabit inland marshes. The most prevalent is probably the muskrat (*Ondatra zibethicus*). This herbivore reproduces rapidly and can attain population densities that decimate the marsh, causing major changes in its character. Like plants, each mammalian species has preferred habitats. For example, Figure 3-15 shows the distribution of muskrats and other mammals on an elevation gradient in a Czech fishpond littoral marsh. Muskrats are found in the most aquatic areas, the water vole in overlapping but higher elevations, and other voles in the relatively terrestrial parts of the reed marsh. Most of the mammals are herbivorous.

Mammals that feed on tree seedlings can potentially influence forest structure. Using animal exclosures, Andersen and Cooper (2000) found that montane voles (*Microtus monatnus*) reduced survivorship of Fremont cottonwood (*Populus fremontii*) seedlings and saplings in a riparian forest along the Green River in Utah. They found that river flow regulation had likely contributed to increased vole pressure in two ways. First, it reduced seasonal snowmelt flooding that normally caused substantial *Microtus* mortality. Second, the reduced flooding encouraged greater herbaceous vegetation which was favorable to voles. The interactions between hydroperiod, plants, and herbivores are rarely understood but can be very important to long-term forest condition.

Birds

Birds, particularly waterfowl, are abundant in freshwater marshes and are one of the reasons these ecosystems are appreciated by so many (Weller, 1999). Many of the birds are herbivorous or omnivorous. Waterfowl are plentiful in almost all wetlands

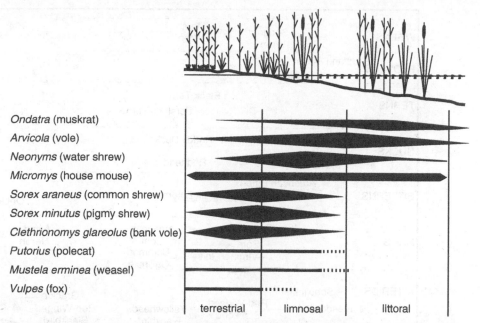

Figure 3-15 Distribution of small mammal populations in the littoral zone of a Czech Republic pond. The three zones are defined as terrestrial, limnosal, and littoral from left to right. (*After Pelikan, 1978*)

probably because of the food richness and the diversity of habitats for nesting and resting. Waterfowl nest in northern freshwater marshes, winter in southern marshes, and rest in other marshes during their migrations. In a typical freshwater marsh, different species distribute themselves along an elevation gradient through marshes according to how well they are adapted to water (see Figure 3-16). In northern marshes, the Loon (*Gavia immer*) usually uses the deeper water of marsh ponds, which may hold fish populations. Grebes (*Podilymbus* sp. and *Podiceps* sp.) prefer marshy areas, especially during the nesting season. Some ducks (dabblers) such as Mallards (*Anas platyrhynchos*) nest in upland sites, feeding along the marsh–water interface and in shallow marsh ponds. Others (diving ducks) such as the Ruddy Duck (*Oxyura jamaicensis*) nest over water and fish by diving. For example, the Black Duck (*Anas rubripes*), one of the most popular ducks for naturalists and hunters alike, uses the emergent marsh as its preferred habitat. The Northern Shoveler (*A. clypeata*), the "whale of the waterfowl," uses its large bill and comblike lamellae to filter plankton. Geese (*Branta canadensis* and *Chen* sp.) and swans (*Cygnus* sp.), the "cattle of the waterfowl," along with Canvasback Ducks (*Aythya valisineria*) and the Wigeon (*Anas americana*), are major marsh herbivores. Wading birds such as the Great Blue Heron (*Ardea herodias*) and the Great Egret (*Casmerodius albus*) usually nest colonially in wetlands and fish along the shallow ponds and streams. The Least Bittern (*Ixobrychus exilis*) builds nests a meter or less above the water in *Typha* or *Scirpus/Schoenoplectus*

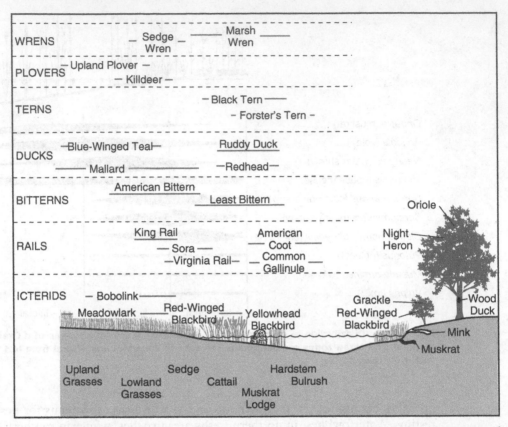

Figure 3-16 Typical distribution of birds across a freshwater marsh from open water edge across shallow water to upland grasses. Placement of muskrat and mink is also illustrated. (*After Weller and Spatcher, 1965*)

stands. Rails live in the whole range of wetlands; many of them are solitary birds that are seldom seen. Marsh Wrens (*Cistohorus palustris*), Virginia Rails (*Rallus limicola*), Soras (*Porzana carolina*), and Swamp sparrows (*Melospiza georgiana*) live amid the dense vegetation of freshwater marshes, often heard but not seen. Song-birds are also abundant in and around marshes. They often nest or perch in adjacent uplands and fly into the marsh to feed. Swallows (*Riparia riparia* and *Stelgidopteryx serripennis*) and swifts are common around freshwater marshes, flying above the marsh, with their mouths ever open to capture emerging insects, often in swarms of dozens or even hundreds of birds.

One of the most conspicuous wetland birds is the blackbird, represented by the Yellow-headed Blackbird (*Xanthocephalus xanthocephalus*) in parts of the midwestern United States and the Red-winged Blackbird (*Agelaius phoeniceus*) in the eastern United States. The Red-winged Blackbird is a very social species and is quite territorial, especially during the nesting season (see Özesmi and Mitsch, 1997).

Fish

One of the most difficult issues about which to generalize is whether freshwater marshes support much fish life or indeed if they should. As a general rule, the deeper the water in the marsh and the more open the system is to large rivers or lakes, the more variety and abundance of fish that can be supported. The positive aspect of freshwater marshes as habitats and nurseries for fish was investigated by Derksen (1989) for a large Manitoba marsh complex and by Stephenson (1990) for Great Lakes marshes. Derksen found extensive use of marshes by northern pike (*Esox lucius*) with emigration from the marsh occurring primarily in the autumn. Stephenson investigated fish utilization for spawning and rearing of young in five marshes connected to Lake Ontario in Ontario, Canada. A total of 36 species of fish were collected, 23 to 27 species per marsh. The spawning adults of 23 species and the young-of-the-year of 31 species were collected in the marshes, indicating the importance of these marshes for fish reproduction in Lake Ontario. Eighty-nine percent of the species encountered were using the marshes for reproduction (Stephenson, 1990).

Common carp (*Cyprinus carpio*) are able to withstand the dramatic seasonal and diel fluctuations of water temperature and dissolved oxygen typical of shallow marshes and are, thus, abundant in many inland wetlands. They affect marsh vegetation by direct grazing, uprooting vegetation while searching for food, and causing severe turbidity in the water column. For these reasons, carp are not considered desirable by many freshwater marsh managers.

Amphibians

Amphibians are an important group of organisms in freshwater marshes, often serving as the link between insect populations and wading birds, mink, raccoons, and some fish in complex food webs. Larval tadpoles, which can be quite abundant in some freshwater marshes, eat small plants and animals and are, in turn, eaten by large fish and wading birds. The adult frogs feast on emerging insects. Even terrestrial toads use freshwater marshes as mating and breeding grounds in the spring. There has been concern about declining amphibian populations; one of the causes that has been suggested has been the loss of wetland habitat. Richter and Azous (1995) investigated the relationships between amphibian richness and variables such as wetland size, vegetation type, presence of competitors and predators, hydrologic characteristics, hydroperiod fluctuations, and land use. The variables that explained the highest correlation with amphibian richness were water-level fluctuations and percentage of the watershed that was urbanized. These data do not explain the exact cause of the loss of amphibians, but urban pollution and stream and hydrological modifications appear to be likely causes.

Porej (2004) and Porej and Hetherington (2005) compared several created and restored wetlands in central Ohio and found that the presence of a shallow-sloped littoral zone, the absence of fish (often caused by flooding limited to seasonal patterns in the wetland), and a high ratio of edge to area of wetlands (optimized where there are many small basins as opposed one large basin of the same area) are among the key physical and biological features that support a diversity of amphibians. The presence of American toads (*Bufo americanus*), northern leopard frogs (*Rana pipiens*), western chorus frogs (*Pseudacris triseriata*), gray tree frogs (*Hyla versicolor*), and smallmouth salamanders (*Ambystoma texanum*) was positively correlated with the presence of a shallow littoral zones in these freshwater marshes and ponds. Porej (2004) also found highest salamander richness in forested wetlands compared to freshwater marshes (natural or created) while frogs and toads had similar richness in forested wetlands and marshes but lower than the richness in newly created marshes (see Table 3-8). He found a strong association between the presence of forest cover within 200 m of freshwater wetlands and amphibian diversity in agricultural landscapes, particularly for spotted salamanders (*Ambystoma maculatum*), Jefferson's salamander complex (*A. jeffersonianum* complex), smallmouth salamanders (*A. texanum*), and wood frogs (*Rana sylvatica*).

Ecosystem Function

Primary Production

The productivity of inland marshes has been reported in a number of studies (see Table 3-9). Estimates are generally quite high, ranging upward from about $1,000 \text{ g m}^{-2} \text{ yr}^{-1}$. Some of the best estimates, which take into account underground production as well as that above ground, come from studies of fishponds in former Czechoslovakia. (These are small artificial lakes and bordering marshes used for fish culture.) These estimates, some indicating values of over $6,000 \text{ g m}^{-2} \text{ yr}^{-1}$, are high compared with most of the North American estimates. The productivity is higher than even the productivity of intensively cultivated farm crops.

The emergent monocotyledons *Phragmites* and *Typha*, two of the dominant plants in freshwater marshes, have high photosynthetic efficiency. For *Typha*, efficiency is highest early in the growing season, gradually decreasing as the season progresses. *Phragmites*, in contrast, has a fairly constant efficiency rate throughout most of the growing season. The efficiencies of conversion by these plants in optimum environments of 4 to 7 percent of photosynthetically active radiation are comparable to those calculated for intensively cultivated crops such as sugar beets, sugarcane, or corn.

Productivity variation is undoubtedly related to a number of factors. For example, Gorham (1974) established the close positive relationship between aboveground biomass and summer temperatures (see Figure 3-17). Innate genetic differences among species accounts for part of the variability. For example, in one study that used the same techniques of measurement (Kvet and Husak, 1978), *Typha angustifolia* production was determined to be double that of *T. latifolia*. *T. angustifolia*, however, is typically found in deeper water than that of the habitats of the other species, and so environmental factors were not identical for purposes of comparison.

The dynamics of underground growth are much less studied than those of aboveground growth. Annuals generally use small amounts of photosynthate to support root growth, whereas species with perennial roots and rhizomes often have root:shoot ratios well in excess of one. This relationship also appears to hold true

Table 3-8 **Occurrence (% of Wetlands Occupied) of Pond-Breeding Amphibians in 54 Natural (Emergent and Forested) and 42 Created Wetlands Located in the Till Plains and Glaciated Plateau Ecoregions of Central Ohio**

Species	Natural emergent wetlands	Natural forested wetlands	Created wetlands
American/Fowler's toad (*Bufo americanus/Bufo fowleri*)	15	20	50
Green frog (*Rana clamitans melanota*)	60	59	74
Northern leopard frog (*R. pipiens*)	74	46	76
American bullfrog (*R. catesbeiana*)	33	26	55
Wood frog (*R. sylvatica*)	0	56	0
Spring peeper (*Pseudacris crucifer*)	87	67	52
Western chorus frog (*P. triseriata*)	27	31	23
Gray treefrog (*Hyla versicolor*)	20	26	48
Blanchard's cricket frog (*Acris crepitans blanchardii*)	0	0	12
Frogs and toads (ave ± st error), species per wetland	**3.2 ± 0.3**	**3.0 ± 0.3**	**3.9 ± 0.3**
Tiger salamander (*Ambystoma tigrinum*)	43	47	5
Spotted salamander (*A. maculatum*)	7	43	5
Smallmouth salamander (*A. texanum*)	15	64	14
Jefferson's salamander complex (*A. jeffersonianum* complex)	8	57	0
Marbled salamander (*A. opacum*)	0	7	0
Red-spotted newt (*Notophthalamus viridescens v.*)	0	22	2
Salamanders (ave ± st error), species per wetland	**1.0 ± 0.3**	**2.4 ± 0.2**	**0.3 ± 0.1**
Total amphibians, species per wetland	**4.2 ± 0.5**	**5.4 ± 0.3**	**4.2 ± 0.3**

Source: Adapted from Porej (2004).

Table 3-9 Selected Primary Production Estimates for Inland Freshwater Marshes

Dominant Species	Location	Net Primary Productivity (g m^{-2} yr^{-1})	Reference
Reeds and Grasses			
Glyceria maxima	Lake, Czech Republic	900–4,300[a]	Kvet and Husak (1978)
Phragmites communis	Lake, Czech Republic	1,000–6,000[a]	Kvet and Husak (1978)
P. communis	Denmark	1,400[a]	Anderson (1976)
Panicum hemitomon	Floating coastal marsh, Louisiana	1,700[b]	Sasser et al. (1982)
Schoenoplectus lacutstris	Lake, Czech Republic	1,600–5,500[a]	Kvet and Husak (1978)
Sparganium eurycarpum	Prairie pothole, Iowa	1,066[b]	van der Valk and Davis (1978)
Typha glauca	Prairie pothole, Iowa	2,297[b]	van der Valk and Davis (1978)
T. latifolia	Oregon	2,040–2,210[a]	McNaughton (1966)
Typha sp.	Lakeside, Wisconsin	3,450[a]	Klopatek (1974)
Sedges and Rushes			
Carex atheroides	Prairie pothole, Iowa	2,858[b]	van der Valk and Davis (1978)
Carex lacustris	Sedge meadow, New York	1,078–1,741[a]	Bernard and Solsky (1977)
Juncus effusus	South Carolina	1,860[a]	Boyd (1971)
Scirpus fluviatilis	Prairie pothol, Iowa	943[a]	van der Valk and Davis (1978)
Broad-Leaved Monocots			
Acorus calamus	Lake, Czech Republic	500–1,100[a]	Kevt and Husak (1978)

[a] Above- and belowground vegetation.
[b] Aboveground vegetation.

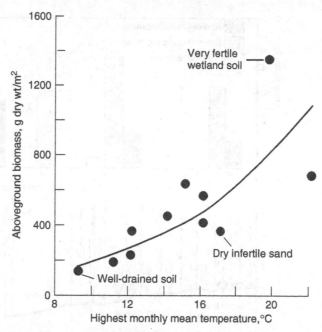

Figure 3-17 Relationship between highest mean monthly temperature and aboveground standing crop of various sedges in freshwater wetlands and uplands. Data points are for wetlands except where noted otherwise. (*After Gorham, 1974*)

for inland freshwater marshes. Perennial species in freshwater marshes generally have more belowground than aboveground biomass (see Figure 3-18). It is interesting to note, however, that, whereas biomass root:shoot ratios are usually greater than one, root-production-to-shoot biomass ratios are generally less than one. Because aboveground production is approximated by aboveground biomass, this latter ratio is an index of the allocation of resources by the plant, and it indicates that less than one-half of the photosynthate is translocated to the roots. The coexistence of large root biomass and relatively small root production suggests that the root system is generally longer lived (i.e., it renews itself more slowly) than the shoot.

Decomposition

With some notable exceptions such as muskrat and geese grazing, herbivory is considered fairly minor in inland marshes where most of the organic production decomposes before entering the detrital food chain. The decomposition process is much the same for all wetlands. Variations stem from the quality and resistance of the decomposing plant material, the temperature, the availability of inorganic nutrients to microbial decomposers, and the flooding regime of the marsh.

Consumers play a significant role in detrital cycles. Most litter decomposition studies in freshwater marshes were done with senesced plant material during the

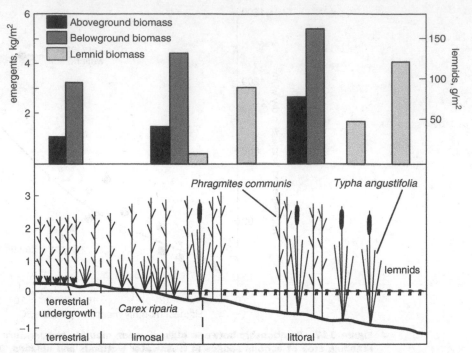

Figure 3-18   Distribution of aboveground and belowground biomass of emergent vegetation and lemnids across a reed bed (*Phragmites*) transect, showing relation to elevation and flooding. Three zones are as in Figure 3-15. (*After Kvet and Husak, 1978*)

winter, and low ($k = 0.002$–0.007 day^{-1}) rates were generally measured. However, in a comparison of the decay of fresh biomass and senesced wetland plant leaves, Nelson et al. (1990a, b) found that samples of freshly harvested wetland plant material (*Typha glauca*) decomposed more than twice as fast ($k = 0.024$ day^{-1}) as naturally senesced material did ($k = 0.011$ day^{-1}). This comparison illustrates the more rapid decomposition that results when animals such as muskrats harvest live plant material.

Muskrats also may play a positive role in the energy flow of a marsh as they harvest aquatic plants and standing detritus for their muskrat mounds. Wainscott et al. (1990) found in culturing experiments that litter from muskrat mounds supports substantially higher densities of microbes than litter from the marsh floor does. They suggested that these muskrat mounds may act like "compost piles" as they accelerate the decomposition and microbial growth that have become familiar to organic gardeners.

Muskrat eat-outs and the resulting open water contribute to structural and biogeochemical heterogeneity within marshes. In an Iowa prairie pothole marsh, Rose and Crumpton (2006) found that as vegetated zones transitioned into open water, there was a predictable decrease in dissolved O_2 and increase in CH_4 concentrations. They suggested that these conditions were regulated by the presence of emergent vegetation and its influence on aerobic and anaerobic metabolism.

Food Webs

Even though food chains begin in the detrital material of freshwater marshes, they develop into detailed webs that are still poorly understood. Benthic communities that feed on detritus form the basis of food for fish and waterfowl in the marshes. DeRoia and Bookhout (1989) found that chironomids made up 89 percent of the diet of Blue-winged Teal (*Anas discors*) and 99 percent of the diet of Green-winged Teal (*A. crecca*) in a Great Lakes marsh. The direct grazing of freshwater marsh vegetation has occasionally been reported in the literature. Crayfish are often important consumers of macrophytes, particularly of submersed aquatic plants, in freshwater marshes. The red swamp crayfish (*Procambarus clarkii*) was shown to have effectively grazed on *Potamogeton pectinatus* in a freshwater marsh in California, where the plant decreased from 70 to 0 percent cover of the marsh while the crayfish population almost doubled (Feminella and Resh, 1989). The direct consumption of marsh plants by geese, muskrats, and other herbivores is common in some parts of the world. "Eat-outs" causing large expanses of open water are the result of the inability of plants to survive after being clipped below the water surface by animals (Middleton, 1999).

In one study that disputed the low-herbivory assumption of inland marshes, deSzalay and Resh (1996) found herbivores in the benthic community of brackish inland marshes in California to represent about 27 percent of the total number of invertebrates collected and to be seasonal. These herbivores fed primarily on filamentous algae and diatoms. Yet herbivory on marsh macrophytes generally remains low except for large-animal grazing from time to time.

Energy Flow

It is not possible to calculate a tight organic energy budget for inland marshes from the information that is available. Several components, however, were estimated for the littoral fishpond system in Czechoslovkia (now Czech Republic) (see Table 3-10). This allows at least some perspective to be developed about the major fluxes of energy. Net organic energy fixed by emergent plants ranges from 1,600 to 16,000 kcal m^{-2} yr^{-1}. (For the purpose of comparison, for a site producing 1,000 g m^{-2} yr^{-1} of biomass per year, the energetic equivalent is about 4,500 kcal m^{-2} yr^{-1}.) Most of this net production is lost through consumer respiration. An early study by Cragg (1961) suggested that microbial respiration in peat was about 1,760 kcal m^{-2} yr^{-1} in a *Juncus* moor. No other estimates appear to be available.

Invertebrates, especially microinvertebrates, play an important role in sediments in the flow of organic energy through the ecosystem. Without doubt, they are important in fragmenting the litter so that it can be more readily attacked by bacteria and fungi. These benthic organisms are also important intermediates in the transfer of energy to higher trophic levels. In Czech fishponds, Dvorak (1978) calculated the average benthic macrofaunal biomass, composed mostly of mollusks and oligochaetes, to be 4,266 g/m^2. They were selectively fed on by fish. In quite a different marsh, a *Juncus* moor, Cragg (1961) estimated the respiration rate of the small soil invertebrates to be about 300 kcal m^{-2} yr^{-1}.

Table 3-10   Estimated Energy Budget for a Fishpond Littoral Freshwater *Phragmites* Marsh in Czech Republic

	kcal m^{-2} yr^{-1}
Producers	
Gross production	4,500–27,000
Respiration	2,900–11,000
Net production	1,600–16,000
Consumers	
Decomposers (bacteria and fungi)	1760
Decomposers (small invertebrates)	300
Invertebrate macrofauna	—
Mammal consumption	232
Bird consumption	20[a]
Mammal production	5
Bird production	1

[a]Assuming production = 5% of consumption.
Source: Based on data from Dykyjova and Kvet (1978).

Pelikan (1978) calculated the energy flow through the mammals of a reedswamp ecosystem. The total energy consumption was 235 kcal m^{-2} yr^{-1}, mostly by herbivores that ingested 220 kcal m^{-2} yr^{-1}. Insectivores ingested 10 kcal m^{-2} yr^{-1}, and carnivores ingested 1 kcal m^{-2} yr^{-1}. This amounted to about 0.55 percent of aboveground and 0.18 percent of belowground plant productivity. Most of the assimilated energy was respired. The total mammal production was only 4.84 kcal m^{-2} yr^{-1}, which amounts to less than 1 g m^{-2} yr^{-1}. It seems evident that the indirect control of plant production by the muskrat is much more significant than the direct flow of energy through this group.

The same reedswamp ecosystem supported an estimated 83 nesting pairs of gulls (*Larus ridibundus*) and 20 pairs of other birds per hectare—about 13 passerines, 3 grebes, and the rest rails and ducks (Hudec and Stastny, 1978). The mean biomass was 44.4 kg ha^{-1} (fresh weight) for the gulls and 11.2 kg ha^{-1} for the remaining species. The production of eggs and young amounted to about 6,088 kcal ha^{-1} for gulls and 3,096 kcal ha^{-1} for the remaining species. If this production were considered to be 5 percent of the total consumption, the total annual flow of organic energy through birds would be about 20 kcal m^{-2}, or roughly 10% of the mammal contribution.

Collectively, these estimates of energy flow through invertebrates, mammals, and birds account for less that 10 percent of net primary production. Most of the rest of the energy used for organic production must be dissipated by microbial respiration, but some organic production is stored as peat, reduced to methane, or exported to adjacent waters. Export has been particularly difficult to measure because so many of these marshes intercept the water table and may lose organic materials through groundwater flows.

Nutrient Budgets

Vegetation traps nutrients in biomass, but the storage of these nutrients is seasonally partitioned in aboveground and belowground stocks. For example, nutrient stocks in the roots and rhizomes of macrophytes are mobilized into the shoots early in the growing season and increase to as much as $4\,g\,P\,m^{-2}$ during the summer. In the fall, some nutrients in the shoots are translocated into the belowground organs before the shoots die, but most of it is lost by leaching and in the litter. C. S. Smith et al. (1988) measured the seasonal changes in several elements of aboveground and belowground parts of *Typha latifolia* from a Wisconsin marsh and found that belowground biomass concentrations of nitrogen, phosphorus, and potassium decreased significantly during the spring, supposedly because of shoot growth, whereas calcium, magnesium, manganese, sodium, and strontium showed little decrease in spring, suggesting that they are not limiting mineral reserves for spring growth.

Studies like those described previously lead to several generalizations about nutrient cycling in freshwater marshes:

- The size of the plant stock of nutrients in freshwater marshes varies widely in contrast to the much more abundant storage of nutrients in marsh soils. More nitrogen and phosphorus are retained in aboveground plant parts in mineral substrate wetlands (freshwater marshes) than in peatlands due to the higher productivity and higher concentrations of nutrients. The aboveground stock of nitrogen ranges from as low as $3\,g\,m^{-2}$ to as high as $29\,g\,m^{-2}$ in freshwater marshes. The nitrogen and phosphorus budgets for a freshwater Scirpus marsh in Wisconsin (see Figure 3-19) show a peak biomass storage of $20.7\,g\,N\,m^{-2}$ and $5.3\,g\,P\,m^{-2}$. This plant storage is small compared to the nutrients that are stored within the root zone of the peat and mineral soils, shown in Figure 3-19 to be $1{,}700\,g\,N\,m^{-2}$ and $12\,g\,P\,m^{-2}$ for total nitrogen and available phosphorus, respectively.

- The biologically inactivated stock of nutrients in plants is only a temporary storage that is released to flooding waters and sediments when the plant shoots die in autumn. Where this occurs, the marsh may retain nutrients during the summer and release them in the winter.

- Nutrients retained in biomass often account for a small portion of nutrients that flow into marshes, and that percentage decreases with increased nutrient input. Shaver and Melillo (1984) illustrated the importance of nutrient availability for the efficiency of nutrient uptake of three freshwater marsh plants: *Carex lacustris, Calamagrostis canadensis*, and *Typha latifolia*. The efficiency of nutrient uptake, defined as the increase in plant nitrogen or phosphorus mass divided by the nitrogen or phosphorus mass available, decreased with increasing nutrient availability, suggesting that plant uptake becomes less important with higher nutrient inputs even though the nutrient content of the plants may increase. The higher concentrations of nutrients in the plants also cause higher concentrations in the litter, which, in turn, can

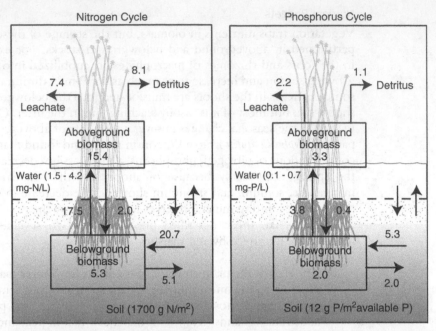

Figure 3-19 Fluxes of nitrogen and phosphorus through a river bulrush (*Scirpus fluviatilis*) stand in Wisconsin. Flows are in g m^{-2} y^{-1} and storages are in g/m^2 of nitrogen and phosphorus, respectively. (*Klopatek, 1978*)

stimulate microbial release. Thus, as more nutrients become available to a freshwater marsh, the marsh becomes more "leaky." Nutrients are lost from the system, and nutrient turnover in the vegetation increases. Even if the uptake rate of nutrients is high in wetlands, much of those nutrients are returned via detrital decomposition to the nutrient pool in the sediments and overlying waters. If wetlands are used for nutrient removal (see Mitsch et al., 2000), then estimates of only 10 to 20 percent of the nutrient inflow going into plant biomass are common.

■ Marsh vegetation often acts as a nutrient pump, taking up nutrients from the soil, translocating them to the shoots, and releasing them on the marsh surface during senescence. The effect of this pumping mechanism may be to mobilize nutrients that have been sequestered in the soil. In some cases, the uptake of nutrients by macrophytes from the sediments is considerably higher than the inflow. Most of this uptake is translocated back to the roots or lost through leaching and shoot senescence, so biomass storage of nutrients is generally low compared to annual inflow.

■ In general, precipitation and dry fall account for less than 10 percent of plant nutrient demands in freshwater marshes. Similarly, groundwater flows are usually small sources of phosphorus, but, in agricultural settings with artificial drainage, nitrate–nitrogen inflow can be high. Surface inflow is usually a major source of phosphorus because of its ability to sorb onto sediments, particularly clay. Considering all of these variables, it is not surprising that each marsh seems to have its own unique nutrient budget. In low-nutrient wetlands such as the Florida Everglades, the marsh system is accustomed to relying primarily on nutrient inflow from precipitation and dry fallout from fires.

Nutrient Limitations

Both nitrogen and phosphorus limitations can be factors in freshwater marsh productivity, and their uptake rates are not independent of each other. Shaver and Melillo (1984) illustrated that their freshwater marsh plants accumulated nutrients at a relatively narrow range and that all plants had a similar "optimum" N:P ratio of approximately 8:1 by mass. Higher nitrogen concentrations in available solutions did increase the N:P ratios of the plants, and low N:P ratios in available solutions decreased the N:P ratios of the plants. Thus, nitrogen and phosphorus uptakes were not independent of each other.

Koerselman and Meuleman (1996), in a study of several wetlands in Europe, found that the N:P ratios in wetland plant tissues were correlated with the N:P supply ratio and that any N:P ratio less than 14:1 suggests nitrogen limitation. This is twice the often-used Redfield ratio (N:P = 7.2) that is used in planktonic systems to indicate relative nutrient limitation. As part of an extensive literature review on temperate North American wetlands, Bedford et al. (1999) found that only marshes were consistently N-limited as indicated by leaf-tissue and soil N:P ratios <14 (although swamps tended to have soils with N:P ratios <14 as well). Based on their leaf tissue N:P ratio, other wetland types (swamps, bogs, fens) tended to be co-limited by N and P or just P-limited based on the thresholds derived by Koerselman and Meuleman (1996).

McJannet et al. (1995) investigated the nitrogen and phosphorus content of 41 freshwater marsh plants after they were grown in excess fertilizer for one growing season. There was a wide range of nitrogen (0.25–2.1 percent N) and phosphorus (0.13–1.1 percent P) that was not related to where the plants came from. However, plants that were from ruderal life histories (i.e., annuals or functional annuals) did have significantly lower nitrogen and phosphorus tissue concentrations than did perennials.

For a Manitoba *Scirpus acutus* marsh, Neill (1990c) found that neither nitrogen nor phosphorus increased net productivity when applied alone but that aboveground biomass nearly doubled when nitrogen and phosphorus were applied together. On the other hand, similar studies of a nearby marsh showed nitrogen limitation, indicating that differences in limiting factors are possible even in the same region (Neill, 1990c). Under conditions in which water levels are more stable such as Louisiana's Gulf Coast, the addition of nitrogen fertilizer at a rate of $10 \, \text{g} \, \text{NH}_4{}^+\text{–N}/\text{m}^2$ caused approximately a 100 percent increase in the growth of *Sagittaria lancifolia* (Delaune and Lindau, 1990).

Experiments by Svengsouk and Mitsch (2001); see Figure 3-20) support the multiple-nutrient limitation of some freshwater marsh plants. Their study investigated

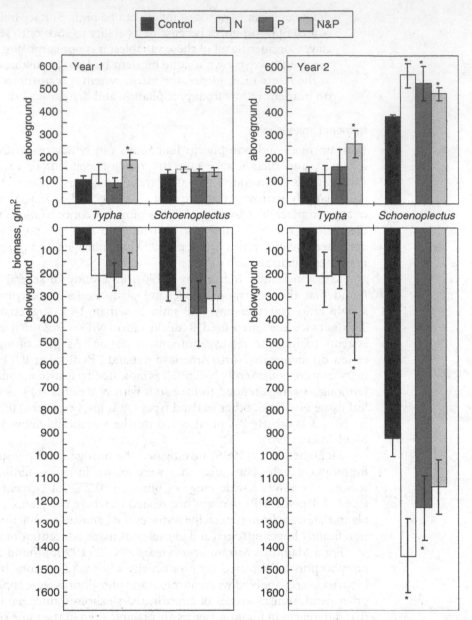

Figure 3-20 Biomass of *Typha* and *Schoenoplectus tabernaemontani* in mesocosms where they were planted together and fertilized with nitrogen (N), phosphorus (P), and nitrogen plus phosphorus (N&P) after one and two growing seasons in experimental mesocosms in Ohio (mean ± std. error; * = significant difference from control at α = 0.05). (*After Svengsouk and Mitsch, 2001*)

the relative limitations of nitrogen and phosphorus in mesocosms planted with both bulrush (*Schoenoplectus tabernaemontani*) and cattail (*Typha* sp.) together. Results suggested that, when both nitrogen and phosphorus are available, *Typha* competed well with *Schoenoplectus*. When only one of the nutrients was in abundance, *Schoenoplectus* did much better than *Typha*.

In contrast to this study, enrichment studies by Craft et al. (1995) on the low-nutrient sawgrass (*Cladium jamaicense*) and other macrophyte communities illustrated that the most important limiting factor in the Florida Everglades appears to be phosphorus. Nitrogen additions had no effect on biomass production, nutrient uptake rates, or nitrogen enrichment of peat. Phosphorus enrichment from agricultural sources caused substantial ecological change in the Everglades, most notably the transition of large areas from sawgrass to cattail (*Typha* spp.). Restoring the Everglades will require reducing P-enriched water into the region. Water quality improvements into the Everglades were implemented in the 1990s and Childers et al. (2003) reexamined four extended transects in 1999 to compare vegetation and soil conditions documented along these transects in 1989. They found that cattails continued to expand from 0.5 to 1.0 km from an inlet to the Loxahatchee National Wildlife Refuge and from 2 to 4 km from a canal inlet at Water Conservation Area 2. Soil P levels also followed this trend and overall species richness declined during this time. It was noted that study plots with higher soil P and mixed sawgrass-cattail in 1989 had since become dominated by cattail suggesting a time lag between shifts in soil nutrient enrichment and *Typha* colonization.

The role that nutrients play in plant productivity and species composition in freshwater marshes is also influenced by hydrologic conditions. Fluctuating water levels may affect nutrient limitations; when nitrogen and phosphorus fertilizers were added to prairie lacustrine marshes over 2 years, nitrogen additions stimulated the productivity of emergent macrophytes in the first year by increasing the *Scolochloa festucacea* biomass more when the soil was flooded and the *Typha latifolia* biomass more in intermediate depths (0–20 cm) than in deep water (20–40 cm) or dry conditions (Neill, 1990a). This response to added nitrogen indicated nitrogen limitation in these marshes. In the second year of the experiment, nitrogen was not as limiting as it was in the first year, and some phosphorus limitation was noted. The species composition in these experiments changed when nitrogen was added, causing a decrease in the *Scolochloa festucacea* biomass and an increase in moist-soil annuals. The phosphorus additions, however, had little effect on species composition (Neill, 1990b).

Recommended Readings

Coburn, E. A. 2004. *Vernal Pools*. McDonald & Woodward Pub. Co., Blacksburg, VA. 426 pp.

Lodge, T. E. 2005. *The Everglades Handbook: Understanding the Ecosystem, 2nd ed.*, CRC Press, Boca Raton, FL. 302 pp.

Messina, M. G., and W. H. Conner, eds. 1998. *Southern Forested Wetlands*. Lewis Publishers, Boca Raton, Florida, 616 pp.

Mitsch, W. J. and J. G. Gosselink. 2007. *Wetlands*, *4th ed.*, John Wiley & Sons, Inc., Hoboken, New Jersey, 582 pp.

van der Valk, A. G. 2006. *The Biology of Freshwater Wetlands*. Oxford University Press, Oxford, UK, 173 pp.

Weller, M.W. 1999. *Wetland Birds*. Cambridge University Press, Cambridge, UK, 271 pp.

Chapter 4

Peatlands

Peatlands include bogs and fens distributed primarily in the cool boreal zones of the world where excess moisture is abundant. Bogs and fens can be formed in several ways, originating either from aquatic systems, as in flowthrough succession or quaking bogs, or from terrestrial systems, as with blanket bogs. Although many types of peatlands are identifiable, classification according to chemical conditions usually defines three types: (1) minerotrophic (true fens), (2) ombrotrophic (raised bogs), and (3) transition (poor fens). Features of many peatlands include acidity caused by cation exchange with mosses, oxidation of sulfur compounds, and organic acids; low nutrients and primary productivity; slow decomposition; adaptive nutrient-cycling pathways; and peat accumulation. Several energy and nutrient budgets have been developed for peatlands, with the 1942 energy budget by Lindeman one of the best known. Peatlands have been shown to be nutrient sinks in long term studies as long as sufficient area is used in the analysis. Also, they are effective sinks for excess atmospheric carbon in undisturbed watersheds.

As defined here, peatlands (see Figure 4-1) include the deep peat deposits of the boreal regions of the world. Bogs and fens, the two major types of peatlands, occur as thick peat deposits in old lake basins or as blankets on the landscape. Many of these lake basins were formed by the last glaciation, and the peatlands are considered to be a late stage of a "filling-in" process. *Bogs* are acid peat deposits with no significant inflow or outflow of surface water or groundwater, and support *acidophilic* (acid-loving) vegetation, particularly mosses. *Fens*, on the other hand, are open peatland systems that generally receive some drainage from surrounding mineral soils and are often covered by grasses, sedges, or reeds. They are in many respects transitional between marshes and bogs. As a successional stage in the development of bogs, fens are important and will be considered in that context here.

Figure 4-1 Northern peatland in Estonia. Photograph by W. J. Mitsch.

Bogs and fens have been studied and described on a worldwide basis more extensively than any other type of freshwater wetland; European and North American ecology literatures are particularly rich in peatland studies. Peatlands have been studied because of their vast area in temperate climates, their unique biota and successional patterns, their economic importance of peat as a fuel and soil conditioner, and, recently, their importance in the global atmospheric carbon balance. Bogs have intrigued and mystified many cultures for centuries because of such discoveries as the Iron Age "bog people" of Scandinavia, who were preserved intact for up to 2,000 years in the nondecomposing peat (see, e.g., Glob, 1969; Coles and Coles, 1989).

Because bogs and other peatlands are ubiquitous in northern Europe and North America, many definitions and words, some unfortunate, that now describe wetlands in general originated from bog terminology. Also, considerable confusion exists in the use of terms such as *bog, fen, swamp, moor, muskeg, heath, mire, marsh, highmoor, lowmoor*, and *peatland* to describe these ecosystems (see Chapter 2 in Mitsch and Gosselink, 2007 for definitions of these terms). The use of the words *peatlands* in general and *bogs* and *fens* in particular will be used in this chapter to include deep peat deposits, mostly of the cold, northern, forested regions of North America and Eurasia. Peat deposits also occur in warm temperate, subtropical, or tropical regions, and we refer briefly to a major example of these, specifically the *pocosins* of the southeastern Coastal Plain of USA.

Geographical Extent

Bogs and fens are distributed in cold temperate climates of high humidity, mostly in the Northern Hemisphere (see Figure 4-2), where precipitation exceeds evapotranspiration, leading to moisture accumulation. There are also some peatlands in the southern hemisphere in southern South America and in New Zealand. But the most extensive areas of bogs and fens occur in Scandinavia, eastern Europe, western Siberia, Alaska, and Canada. Major areas where a very large percentage of the landscape is peatland include the Hudson Bay lowlands in Canada, the Fennoscandian Shield in northern Europe, and the western Siberian lowland around the Ob and Irtysh Rivers.

Estimates range from 2.4 to 5.8 million km^2 for northern boreal and subarctic peatlands (Kivinen and Pakarinen, 1981; Armentano and Menges, 1986; Matthews and Fung, 1987; Aselmann and Crutzen, 1989; Gorham, 1991; Bleuten et al., 2006). There was some convergence on recent estimates to suggest that there are about 3.5 million km^2 of peatlands in the world (Bridgham et al., 2001; Strack, 2008) although Bleuten et al. (2006) used a previously unpublished estimate of 3.1 million km^2 of peatlands in Asian Russia alone to estimate a world total of 5.8 million km^2. Specific country or regional estimates of peatland extent include 1.6 million km^2 for the former Soviet Union (Botch et al., 1995) including 900,000 km^2 in the Western Siberian lowlands (Kremenetski et al., 2003), 1.1 to 1.2 million km^2 for Canada, 220,000 km^2 for Fennoscania (Gorham, 1991), and 130,000 km^2 for Alaska (Bridgham et al., 2001).

Some peatlands, not illustrated in Figure 4-2, are found in the Southern Hemisphere in southern South America and in New Zealand, but the size of these peatlands collectively is small compared to those in the Northern Hemisphere. For example, the New Zealand Land Resource Inventory (Cromarty and Scott, 1996) lists 3,113 km^2 of wetlands in the entire country, many of which are peatlands. There are 43,900 ha of *pakihi* (shallow-peat heathland) and another 35,600 ha of forest–pakiha associations, some of which support sphagnum moss (Buxton et al., 1996). Sanders and Winterbourn (1993) estimate that there may be as much as 8,000 ha on South Island where *Sphagnum* moss is being harvested for horticultural purposes. Otherwise, raised bogs in New Zealand are not characterized by *Sphagnum* moss or ericaceous

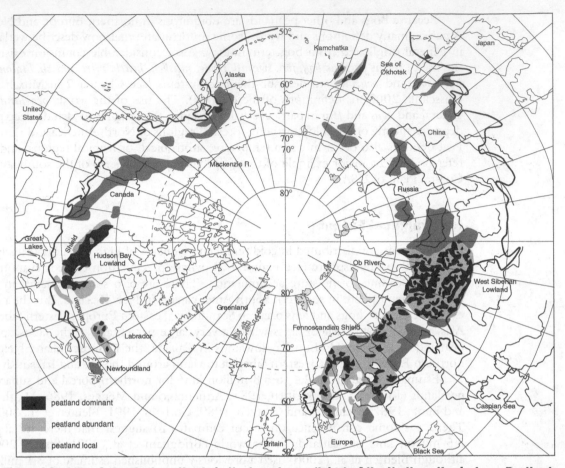

Figure 4-2 Area of abundant peatlands in the boreal zone (taiga) of the Northern Hemisphere. Peatlands are associated with boreal regions and their subalpine equivalents in mountainous regions. South of the tree line (solid line), is a transitional subalpine and woodland tundra zone. (*After Wieder et al., 2006*)

species common in the Northern Hemisphere but by rush-like plants co-dominated by restiad (coming from the family Restionaceae) bog species. In the Hamilton region of North Island, there were originally 50,000 ha of these peatlands, but less than 140 ha remain; another 9,200-ha bog is further to the north (Clarkson et al., 1999).

The distribution of peatlands in North America is shown in Figure 4-3. Canada has approximately 1.10 million km² of peatlands (Zoltai, 1988). Combining this with an estimated 0.55 million km² of peatlands in the United States, including Alaska (see Table 3-1), the 1.65 million km² of northern peatlands in North America represent at least one-third of the world's peatlands. These peatlands dip into the conterminous United States from northern Minnesota to northern Maine. In the northeast United States, bogs are found in Maine, New York, and Vermont. South of the peatlands there

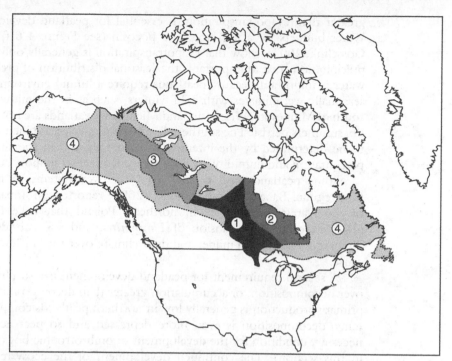

Figure 4-3 Major boreal peatlands of North America including (1) glacial Lake Agassiz region, (2) Hudson Bay lowlands, (3) Great Bear/Great Slave Lake region, and (4) other boreal peatlands including the interior of Alaska. (*After Glaser, 1987*)

tend to be small depressional systems that receive at least some nutrient inflow. In the United States, bogs usually develop in basins scoured out by the Pleistocene glaciers.

Although predominantly a northern phenomenon, peats can accumulate wherever drainage is impeded and anoxic conditions predominate regardless of temperature. Bogs are found as far south as Illinois, Indiana, and Ohio in the north-central United States and are fairly common in the unglaciated Appalachian Mountains in West Virginia. The Middle Atlantic Coastal Plain supports an expansive area of poorly drained peatlands, called *pocosins*, that are similar to more northern peatlands in that they are nutrient poor and dominated by evergreen woody plants such as bilberry, whortleberry, cranberry, heather, and Labrador tea belonging to the Ericaceae or heath family. Pocosins once covered 1.2 million ha, 70 percent of which were in North Carolina. Thirty-three percent of the pocosins in North Carolina have been destroyed (Richardson, 2003).

Hydrology and Peatland Development

Two primary processes necessary for peatland development are a positive water balance and peat accumulation. First, a positive water balance, meaning that precipitation is greater than evapotranspiration, is essential for peatland development and survival. Water budgets for a fen, bog, and pocosin (see Figure 4-6 f, h, i, in Mitsch and Gosselink, 2007), show that evapotranspiration is generally only 50 to 70 percent of precipitation. Just as important, the seasonal distribution of precipitation and excess water is important because peatlands require a humid environment year-around. In seasonally wet climates with cold winters, such as in the midwestern United States south of Minnesota, Wisconsin, and Michigan, peatlands are not common where hot, dry summers persist. The southern limit to bog species and, hence, to bogs is thought to be determined by the intensity of solar radiation in the summer months when precipitation and humidity are otherwise adequate to support bogs.

Some peatlands also depend on local river systems to maintain their moisture regime. Banaszuk and Kamocki (2008) reported on impaired fluvial peatlands of the Narew River valley in northeast Poland that have been impacted by a lower water table, expansion of *Phragmites*, and soil subsidence. Reduced river discharges, linked to a milder and drier climate over the last few decades, may be the cause.

A second requirement for peatland development is a surplus of peat production over decomposition, or accumulation greater than decomposition $(A > D)$. Although primary production is generally low in northern peatlands compared to other ecosystems, decomposition is even more depressed and so peat accumulates. This is a necessary condition for the development of ombrotrophic bogs (see description later in this section). The continued development of the ecosystem is directly related to the amount of surplus water and peat. For example, in a cool, moist maritime climate, peatlands can develop over almost any substrate, even on hill slopes. In contrast, in warm climates where both evapotranspiration and decomposition are elevated, ombrotrophic peatlands seldom develop even when a precipitation surplus occurs. Once formed, a bog is remarkably resistant to conditions that alter the water balance and peat accumulation. The perched water table, the water-holding capacity of the peat, and its low pH create a microclimate that is stable under fairly wide environmental fluctuations.

Given the conditions of water surplus and peat accumulation, peatlands develop through *terrestrialization* (the infilling of shallow lakes) or *paludification* (the blanketing of terrestrial ecosystems by overgrowth of peatland vegetation). Three major bog formation processes are commonly seen:

- **Quaking bog succession:** This is the classical process of terrestrialization, as described in most introductory botany or limnology courses. Bog development in some lake basins involves the filling in of the basin from the surface, creating a *quaking bog* (or *Schwingmoor* in German; Figure 4-4). Plant cover, only partially rooted in the basin bottom or floating like a raft, gradually develops from the edges toward the middle of the lake. A mat of reeds, sedges, grasses, and other herbaceous plants develops along the leading edge of a floating mat of peat that is soon consolidated and dominated by *Sphagnum* and other bog flora. The mat has all of the characteristics of a *raised bog* except

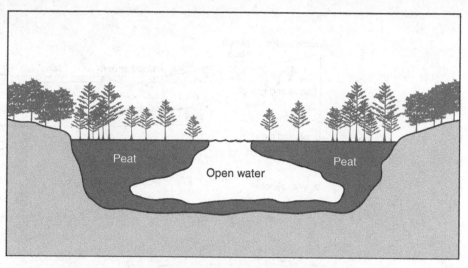

Figure 4-4   Typical profile of a quaking bog.

hydrologic isolation. The older peat is often colonized by shrubs and then forest trees such as pine, tamarack, and spruce, which form uniform concentric rings around the advancing floating mat.

These peatlands develop only in small lakes that have little wave action; they receive their name from the quaking of the entire surface that can be caused by walking on the floating mat. After peat accumulates above the water table, isolating the *Sphagnum*-dominated flora from their nutrient supply, the bog becomes increasingly nutrient poor. The development of a perched water table also isolates the peatland from groundwater and nutrient renewal. The result is a classic concentric, or excentric, ombrotrophic raised bog (see the discussion in the section "Classification of Peatlands," which follows).

The hydrology of raised bogs has been investigated and found to be more complicated than originally thought, particularly for bogs that are on the edge of the boreal zone. Studies in the Lake Agassiz region of Minnesota showed that bogs and fens are part of a regional hydrology, with fens receiving groundwater and raised bogs generally recharging groundwater (see Figure 4-5a). Raised bogs are normally assumed to be disconnected from groundwater and fed only by precipitation. In a normal wet climate, this pattern of bog hydrology is true as a downward flow of excess precipitation deflects upwardly moving groundwater from mineral soil well below the surface (see Figure 4-5b). This accelerates peat accumulation, which, in turn, maintains the peat and, hence, hydrologic mound in the landscape. During droughts, which can be frequent events in peatlands on the edge of the boreal region, groundwater can move upward to within 1 to 2 m of the peat surface (see Figure 4-5c) and dramatically influence the peatland chemistry.

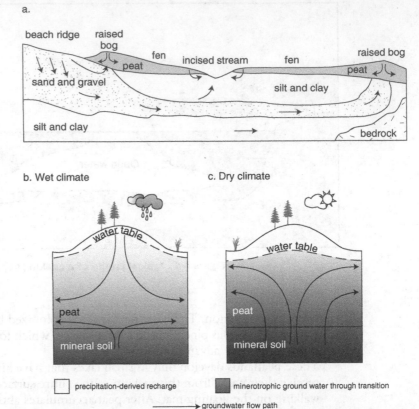

Figure 4-5 a. Regional linkages between groundwater and raised bogs in the Lake Agassiz region of Minnesota (area is approximately 10 km long and 30 m thick). Detailed patterns of subsurface hydrology in the raised bogs are illustrated for b. wet climate and c. dry climate. During wet periods, precipitation-derived recharge maintains a head that flushes mineral-rich groundwater from the peat. During droughts, the water mound drops and mineral-rich groundwater can move upward into the raised-bog peat. (*After Siegel et al., 1995; Glaser et al., 1997a*)

- **Paludification:** A second pattern of bog evolution occurs when blanket bogs exceed basin boundaries and encroach on formerly dry land. This process of paludification can be brought about by climatic change, geomorphological change, beaver dams, logging of forests, or the natural advancement of a peatland. Often, the lower layers of peat compress and become impermeable, causing a perched water table near the surface of what was formerly mineral soil. This causes wet and acid conditions that kill or stunt trees and allow only ombrotrophic bog species to exist. In some situations, the progression from forest to bog can take place in only a few generations of trees (Heilman, 1968).

- **Flowthrough succession:** Intermediate between terrestrialization and paludification is flowthrough succession (also termed *topogenous development*), in which the development of peatland modifies the pattern of surface water flow. The bog may develop from a lake basin that originally had continuous inflow and outflow of surface and groundwater. As the peat continues to build, the major inflow of water may be diverted and areas may develop that become inundated only during high rainfall. In the final stage, the bog remains above the groundwater level and becomes a true ombrotrophic bog.

Classification of Peatlands

Peatlands develop within a complex interaction of climate, hydrology, topography, chemistry, and vegetation development (succession). Because the physical and biotic processes that form peatlands are complex and differ somewhat from region to region, many different classification systems have been proposed over the past century (see Table 4-1). Classification schemes have been based on at least seven features: (1) floristic, (2) vegetation structure, (3) geomorphology (succession or development), (4) hydrology, (5) chemistry, (6) stratigraphy, and (7) peat characteristics. The last is used primarily for economic exploitation purposes. The other six are closely interrelated, leading to classification schemes that combine several natural features.

Landscape Classification

The developmental processes described previously determine large-scale patterns of landform development as summarized in Figure 4-6.

- **Raised bogs:** These are peat deposits that fill entire basins, are raised above groundwater levels, and receive their major inputs of nutrients from precipitation. These bogs are primarily found in the boreal and northern deciduous biomes. When a concentric pattern of pools and peat communities forms around the most elevated part of the bog, the bog is called a *concentric domed bog* (see Figure 4-6a). Bogs that form from previously separate basins on sloping land and form elongated hummocks and pools aligned perpendicular to the slope are called *excentric raised bogs* (see Figure 4-6b). In Europe, the former are found near the Baltic Sea, and the latter are found primarily in the North Karelian region of Finland.

- **Aapa peatlands:** These wetlands (see Figures 4-6c and 4-7), also called *string bogs* and *patterned fens*, are found throughout the boreal region, often north of the raised bog region. The dominant feature of these wetlands is the long, narrow alignment of the higher peat hummocks (*strings*) that form ridges perpendicular to the slope of the peatland and are separated by deep pools (*flarks* in Swedish). In appearance, they resemble a hillside of terraced rice fields. The strings and flarks develop perpendicular to the direction of the water flow. The pattern begins as a series of scattered pools on the down slope, wetter edge of the water track. These pools gradually coalesce into linear flarks. Peat accumulation in the adjacent strings and the increasing

Table 4-1 Historical Classification Schemes for Peatlands

Principal Basis for Classification	Mineral-influenced Peatlands	Transition Peatlands	Precipitation-dominated Peatlands	Reference
Topography	Fen		Bog or raised bog	General use
	Niedermoore (low moor)	Übergangsmoore	Hochmoore (high moor)	Weber (1907)
Hydrology	Geogenous Limnogenous Topogenous Soligenous		Ombrogenous	von Post and Granlund (1926), Sjörs (1948), Du Rietz (1949), Damman (1986)
	Rheophilous	Transition	Ombrophilous	Kulczyński (1949)
	Soligenous		Ombrogenous	Walter (1973)
	Minerogenous		Ombrogenous	Warner and Rubec (1997)
Water chemistry	Rich fen	Poor fen	Bog	General use; Sjörs (1948)
	Minerotrophic	Mesotrophic	Ombrotrophic	Moore and Bellamy (1974)
	Rheotrophic		Ombrotrophic	Moore and Bellamy (1974)
Nutrition	Nährstoffreichere	Mittelreiche	Nährstoffarme	Weber (1907)
	Eutrophic	Mesotrophic	Oligotrophic	Weber (1907), Pjavchenko (1982)
Vegetation	Emergent or forested fen	Transitional	Moss–lichen or forested bog	Cowardin et al. (1979), Gorham and Janssens (1992)

Source: Adapted from Bridgham et al. (1996).

impermeability of decomposing peat in the flarks accentuate the pattern. Within the large water tracks, tree islands appear to be remnants of continuous swamp forests that were replaced by sedge lawns in the expanding water tracks.

- **Paalsa bogs:** These bogs, found in the southern limit of the tundra biome, are large plateaus of peat (20–100 m in breadth and length and 3 m high) generally underlain by frozen peat and silt (see Figure 4-6d). The peat acts like an insulating blanket, actually keeping the ground ice from thawing and allowing the southernmost appearance of the discontinuous permafrost. In Canada, as much as 40 percent of the land area is influenced by cyrogenic factors. When peat overlies frozen sediments, it influences the pattern of the landscape. Many distinctive forms are similar to European aapa and paalsa peatlands but are embedded in a continuous peat-covered landscape.

- **Blanket bogs:** These wetlands (see Figure 4-6e) are common along the northwestern coast of Europe and throughout the British Isles and a result of

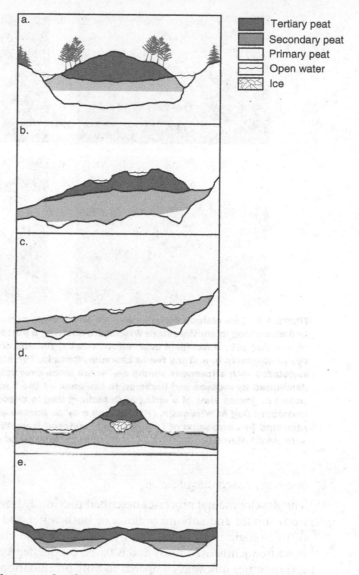

Figure 4-6 Diagrams of major peatland landscape types, including a. raised bog (concentric), b. excentric raised bog, c. aapa peatland (string bogs and patterned fens), d. paalsa bog, and e. blanket bog.

paludification described above. The favorable humid Atlantic climate allows the peat literally to "blanket" very large areas far from the site of the original peat accumulation. Peat in these areas can generally advance on slopes of up to 18 percent; extremes of 25 percent have been noted on slopes covered by blanket bogs in western Ireland.

Figure 4-7 **Two oblique aerial images of string bogs in North America: a. aerial photo of Cedarburg Bog in southwestern Wisconsin, showing a pattern of parallel peat ridges (strings) alternating with water-filled depressions (flarks) running diagonally across the lower half of the photograph; b. a string fen in Labrador, Canada. The strings stand out because they are vegetated with ericaceous shrubs and scrub trees over sphagnum moss, whereas flarks are dominated by mosses and herbs or, in the case of the Canadian site, extensive standing water; c. ground view of a string at Cedarburg Bog in Wisconsin; d. ground view of a flark at Cedarburg Bog in Wisconsin. (*Photograph a by G. Guntenspergen; photograph b by D. Wells, reprinted by permission of C. Rubec and reprinted from Mitsch et al., 1994, p. 30, Fig. 30, with permission from Elsevier Science; photographs c and d by W. J. Mitsch.*)**

Chemistry-Based Classification

The developmental processes described previously lead to increasing isolation of bogs from surface and subsurface flows of both water and mineral nutrients. The degree of hydrologic isolation of mires leads to a simple classification that is probably the most frequently used today and is based on the degree to which the peatland receives groundwater inflow as compared to only precipitation:

- **Minerotrophic peatlands:** These are true fens that receive water that has passed through mineral soil. These peatlands generally have a high groundwater level and occupy a low point of relief in a basin. They are also referred to as *rheotrophic peatlands* and *rich fens* in general use.

- **Mesotrophic peatlands:** These peatlands are intermediate between mineral-nourished (minerotrophic) and precipitation-dominated (ombrotrophic) peatlands. Another term used frequently for this class is *transitional peatlands* or *poor fens*.

- **Ombrotrophic peatlands:** These are the true raised bogs that have developed peat layers higher than their surroundings and that receive nutrients and other minerals exclusively by precipitation.

Another "trophic" classification of peatlands, found in older European literature (Weber, 1907) and originally developed to classify peatlands and not lakes (Hutchinson, 1973), is the three-level trophic classification now familiar to limnologists:

- **Eutrophic peatlands:** Nutrient-rich peatlands described by Weber (1907) as *Nährstoffreichere* (eutrophe).
- **Mesotrophic peatlands:** Same as before, described by Weber (1907) as *Mittelreiche* (mesotrophe).
- **Oligotrophic peatlands:** Nutrient-poor peatlands, described by Weber (1907) as *Nährstoffarme* (oligotrophe).

Hutchinson (1973) suggests that the process of peatland development could be called *oligotrophication*. The terms *eutrophic* and *oligotrophic* were only applied to lakes and their current limnological use a decade later by Naumann (1919). Russian scientists such as Pjavchenko (1982) and Bazilevich and Tishkov (1982) continued to use this nomenclature well into the 1980s.

Bridgham et al. (1996) argued for caution in the use of the "-trophic" suffix for classifying peatlands because the classic peatland gradient from minerotrophic to ombrotrophic, characterized by surface water chemistry such as pH, conductivity, and alkalinity, does not necessarily correlate with the eutrophic to oligotrophic gradient, which is defined in terms of nutrient (e.g., nitrogen, phosphorus, and potassium) availability. In fact, one study (Bridgham et al., 1998) found evidence to suggest that there was higher phosphorus availability in bogs and higher nitrogen availability in fens. In other words, a strict correlation between measures of dissolved minerals and available nutrients has never been established. They suggested resurrecting the terms eutrophic and oligotrophic, which are rarely seen in peatland literature today, because they clearly refer to nutrients and not to other minerals. Such a resurrection has not occurred.

Hydrology-Based Classification

Terms such as *soligenous* and *ombrogenous* actually refer to the hydrological and topographic origins of the peatlands and not to the mineral conditions of the inflowing water. A true hydrologic classification of peatlands based on the following categories is illustrated in Figure 4-8:

1. **Ombrogenous peatlands:** Open only to precipitation
2. **Geogenous peatlands:** Open to outside hydrologic flows other than precipitation:
   a. *Limnogenous peatlands:* Develop along slow-flowing streams or lakes
   b. *Topogenous peatlands:* Develop in topographic depressions with at least some regional groundwater flow
   c. *Soligenous peatlands:* Develop with regional interflow and surface runoff

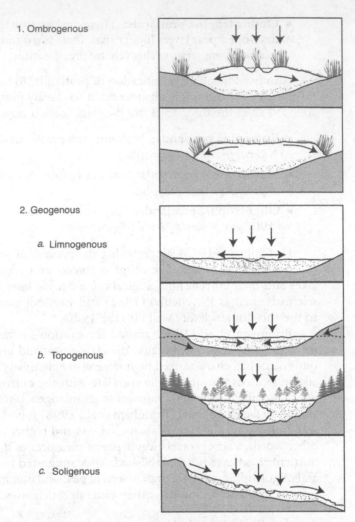

1. Ombrogenous

2. Geogenous

   a. Limnogenous

   b. Topogenous

   c. Soligenous

Figure 4-8 Classification of peatlands based on hydrology. Two major categories are geogenous peatlands, which are open to surface and groundwater flow, and ombrogenous peatlands, which only receive precipitation. (*After Damman, 1986*)

One of the more complete classifications developed for wetlands in general and peatlands in particular is the Canadian Wetland Classification System (Warner and Rubec, 1997). This classification uses *minerotrophic* and *ombrotrophic* in its water chemistry classification and *minerogenous* and *ombrogenous* in its hydrological classification. Their simple classification of peatlands is:

- **Bog:** Peatland receiving water exclusively from precipitation and not influenced by groundwater; sphagnum-dominated vegetation.

- **Fen:** Peatland receiving water rich in dissolved minerals; vegetation cover composed dominantly of graminoid species and brown mosses.

- **Swamp:** Peatland dominated by trees, shrubs, and forbs; waters rich in dissolved minerals.

Biogeochemistry

Soil and water chemistry are among the most important factors in the development and structure of the peatland ecosystems. Factors such as pH, mineral concentration, available nutrients, and cation exchange capacity influence the vegetation types and their productivity. Conversely, the plant communities influence the chemical properties of the soil water. In few wetland types is this interdependence so apparent as in northern peatlands. The major features of peatland biogeochemistry are discussed here.

Acidity and Exchangeable Cations

The pH of peatlands generally decreases as the organic content increases with the development from a minerotrophic fen to an ombrotrophic bog (see Figure 4-9). Fens are dominated by minerals from surrounding soils whereas bogs rely on a sparse

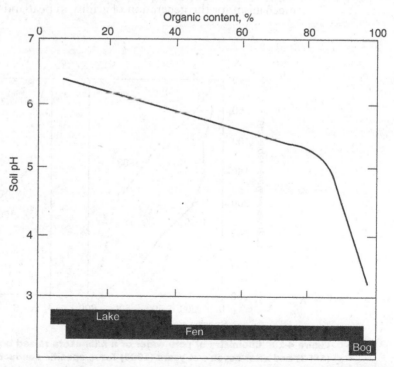

Figure 4-9   Soil pH as a function of organic content of peat soil. (*After Gorham, 1967*)

supply of minerals from precipitation. Therefore, as a fen develops into a bog, the supply of metallic cations (Ca^{2+}, Mg^{2+}, Na^+, K^+) drops sharply. At the same time, as the organic content of the peat increases because of the slowing of the decomposition rate, the capacity of the soil to adsorb and exchange cations increases. These changes lead to the domination by hydrogen ions, and the pH falls sharply. Fens, on the other hand, can range from slightly acidic (poor fens) to strongly alkaline (rich fens) depending on groundwater flow rate and chemistry (Bedford and Godwin, 2003). Gorham (1967) found that bogs in the English Lake District had a pH range of 3.8 to 4.4 compared to non-calcareous fens, which had a pH range of 4.8 to 6.0. The Russian scientist Pjavchenko (1982) assigned a pH range of 2.6 to 3.3 to oligotrophic bogs and a range of 4.1 to 4.8 to mesotrophic bogs; a pH greater than 4.8 defined a eutrophic (minerotrophic) fen.

As little as 10 percent of the water supply from groundwater may change the pH of a bog from 3.6 to 6.8, that is, from an ombrotrophic bog to a minerotrophic rich fen. In an upper peat of a Minnesota raised bog, pH and conductivity, both indicators of mineral groundwater, increased dramatically in a drought year compared to a wet year (see Figure 4-10).

The causes of bog acidity are not entirely clear but five causes are usually cited for the low pH:

- **Cation exchange by Sphagnum:** This may be the most important mechanism for the generation of acidity in peatlands. There is a direct

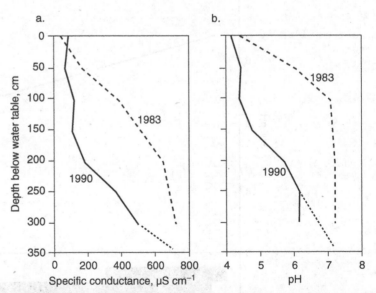

Figure 4-10 Chemistry of pore water of a Minnesota raised bog after a period of drought (1983) and after seven wet years (1990) for a. specific conductance and b. pH. (*After Siegel et al., 1995*)

relationship between pH and the exchangeable hydrogen in peat, presumably the result of the metabolic activity of the plants. *Sphagnum* peats have high exchangeable hydrogen and, consequently, a lower pH than sedge-dominated peats.

- **Oxidation of sulfur compounds to sulfuric acid:** Organic sulfur reserves in peat may be oxidized to acidic compounds (see Mitsch and Gosselink, 2007, Chapter 5).

- **Atmospheric deposition:** Sulfur deposition is a significant source of acidity, depending on the oxidation state of the sulfur and the location of the bog. Acid sources in precipitation and dry deposition are usually small except close to sources of atmospheric pollution.

- **Biological uptake of cations by plants:** Ions in the peat water are concentrated by evaporation and are differentially absorbed by the mosses. This affects acidity, for example, by the uptake of cations that are exchanged with plant hydrogen ions to maintain the charge balance.

- **Buildup of organic acids by decomposition:** Gorham et al. (1984) presented evidence supporting this source of bog acidity. Organic acids help buffer the system against the alkalinity of metallic cations brought in by rainfall and local runoff.

A detailed hydrogen budget constructed for a Minnesota bog complex implicated nutrient uptake as a major source of acidity (see Table 4-2). About 15 percent of this represents ion exchange on the cell walls of *Sphagnum*. Most of this acidity is neutralized by the release of cations during decomposition. Most of the rest of the acidity is generated by organic acid production from fulvic and other acids that result from the incomplete oxidation of organic matter and that buffer the pH of bogs throughout the world at a value of about 4. In addition to decomposition, the major

Table 4-2 Acidity Balance for a Minnesota Bog Complex

Sources	Acidity (meq m^{-2} yr^{-1})
Wet and dry deposition	-0.20 ± 10.7
Upland runoff	-44.3 ± 18.6
Nutrient uptake	827 ± 248
Organic acid production	263 ± 50
Total	1,044
Sinks	
Denitrification	12.2
Decomposition	784
Weathering	76
Outflow	142 ± 50
Total	1,044

Source: Adapted from Urban et al. (1985).

source of alkalinity to neutralize the acids, the weathering of iron and aluminum and runoff, are major processes.

Limiting Nutrients

Bogs are exceedingly deficient in available plant nutrients; fens that contain groundwater and surface water sources generally have considerably more nutrients. The paucity of nutrients in bogs leads to two significant results, which are discussed in more detail later in this chapter: (1) The productivity of nutrient-poor bogs is lower than that of nutrient-rich fens; and (2) the characteristic plants, animals, and microbes have many special adaptations to the low-nutrient conditions. Many studies have attempted to find the ultimate limiting factor for bog primary productivity, which may be a complex and academic question because all available nutrients are in short supply and the growing season is short and cool. Although calcium and potassium have been shown to be limiting, nitrogen and phosphorus are the major limiting chemicals in bog and fen productivity. When these nutrients are added in significant amounts to peatlands, major vegetation shifts occur; with management such as mowing, the limiting factor can change from nitrogen to phosphorus. Bog formation in its latter stages is essentially limited to nutrients brought in by precipitation. The effects on peatlands of increased atmospheric sources of nitrogen throughout the developed world due to fossil fuel burning has yet to be adequately assessed.

Vegetation

Bogs can be simple sphagnum moss peatlands, sphagnum–sedge peatlands, sphagnum–shrub peatlands, bog forests, or any number or combination of acidophilic plants. Mosses, primarily those of the genus *Sphagnum*, are the most important peat-building plants in bogs throughout their geographical range. Mosses grow in cushionlike, spongy mats; water content is high, with water sometimes held higher than it normally would be held by capillary action. *Sphagnum* grows shoots actively only in the surface layers (at a rate of about 1–10 cm annually); the lower layers die off and convert to peat.

In North American peatlands, *Sphagnum* often grows in association with cotton grass (*Eriophorum vaginatum*), various sedges (*Carex* spp.), and certain ericaceous shrubs such as heather (*Calluna vulgaris*), leatherleaf (*Chamaedaphne calyculata*), cranberry and blueberry (*Vaccinium* spp.), and Labrador tea (*Ledum palustre*). Trees such as pine (*Pinus sylvestris*), crowberry (*Empetrum* spp.), spruce (*Picea* spp.), and tamarack (*Larix* spp.) are often found in bogs as stunted individuals that may be scarcely 1 m high yet several hundred years old. Fens in the United States tend to be dominated by a diverse community of plants that are distinct from boreal peatlands and typically include bryophytes, sedges (*Carex* and others genera of Cyperaceae), dicotyledonous herbs, and grasses (Amon et al., 2002; Bedford and Godwin, 2003).

Vegetation Patterns in a Minnesota Peatland

Heinselman (1970) described seven vegetation associations in the Lake Agassiz peatlands of northern Minnesota that are typical of many of those in North America. These occur in an intricate mosaic across the landscape, reflecting the topography, chemistry, and previous history of the site. The seven vegetation zones correspond closely to the underlying peat and to the present nutrient status of the site. The major zones are as follows:

- **Rich swamp forest:** These forested wetlands form narrow bands in very wet sites around the perimeter of peatlands. The canopy is dominated by northern red cedar (*Thuja occidentalis*), but there are also some species of ash (*Fraxinus* spp.), tamarack, and spruce. A shrub layer of alder, *Alnus rugosa*, is often present, as are hummocks of *Sphagnum* moss.

- **Poor swamp forest:** These swamps, occurring downslope of the rich swamp forests, are nutrient-poor ecosystems and are the most common peatland type in the Lake Agassiz region. Tamarack is usually the dominant canopy tree, with bog birch (*Betula pumila*) and leatherleaf in the understory and *Sphagnum* forming hummocks that are 30 to 60 cm high.

- **Cedar string bog and fen complex:** This is similar to zone 2 except that trees'-edge fens alternate with cedar (*Thuja occidentalis*) on the bog ridges (strings) and treeless sedge (mostly *Carex*) in hollows (flarks) between the ridges.

- **Larch string bog and fen:** In this type of string bog, similar to zones 2 and 3, tamarack (*Larix*) dominates the bog ridges.

- **Black spruce–feathermoss forest:** This type is a mature black spruce (*Picea mariana*) forest that also contains a carpet of feathermoss (*Pleurozium*) and other mosses. The trees are tall, dense, and even aged. This peatland occurs near the margins of ombrotrophic bogs and generally does not have standing water.

- **Sphagnum–black spruce–leatherleaf bog forest:** This is a widespread wetland type in northern North America. Stunted black spruce is the only tree, and there is a heavy shrub layer of leatherleaf, laurel (*Kalmia* spp.), and Labrador tea growing in large "pillows" of *Sphagnum* moss between spruce patches. This association is found in convex relief and is isolated from mineral-bearing water.

- **Sphagnum–leatherleaf–kalmia–spruce heath:** A continuous blanket of *Sphagnum* moss is the most conspicuous feature; a low

(*continues*)

(continued)

shrub layer and stunted trees (usually black spruce) are present in five to ten percent of the area. The last two zones occur on a raised bog.

In the water chemistry classification presented earlier in the chapter, the first four zones would be classified as minerotrophic, zone 5 as transitional, zone 6 as semi-ombrotrophic, and zone 7 as ombrotrophic.

Although *Sphagnum* species are the characteristic peat-forming ground cover of bogs, as sedges are of poor fens, there is a considerable overlap of species along the chemical gradient from mineral poor to mineral rich and from low pH to high pH. In a direct gradient analysis of vascular plants found in both bogs and fens in northern Minnesota (see Figure 4-11), the sedges *Carex oligosperma* and *Eriophorum spissum* decrease in cover abundance with mineral enrichment of the peat, whereas tamarack (*Larix laricina*) increases in abundance. Black spruce and the ericaceous

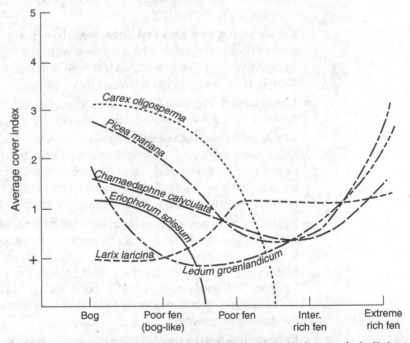

Figure 4-11 Gradient analysis of the major vascular plants that occur in both bogs and fens in Minnesota. Ordinate from Braun–Blanquet scale. Ranges in water chemistry are bog (pH<4.2; Ca^{2+}<2 mg/L), poor fen (bog-like) (pH = 4.1–4.6; Ca^{2+} = 1.5–5.5), poor fen (pH = 4.1–5.8, Ca^{2+}<10 mg/L), intermediate rich fen (pH = 5.8–6.7; Ca^{2+} = 10–32 mg/L), extremely rich fen (pH>6.7; Ca^{2+}>30 mg/L). (*After Glaser, 1987*)

shrubs Labrador tea and leatherleaf, however, show dual peaks, indicating that their distribution is not controlled by mineral water chemistry but by another gradient such as water level or possibly nitrogen or phosphorus availability.

Nicholson et al. (1996) investigated climatic and ecological gradients and how they affected bryophyte distribution in the Mackenzie River basin in northwestern Canada. They found that the most important variables that explained bryophyte species distributions were water chemistry (Mg^{2+}, Ca^{2+}, H^+), height above the water table, precipitation, and annual temperature. As a result of examining these gradients, seven peatland groups were clustered from the original 82 sites in the basin: (1) poor fens, (2) peat plateaus with thermokarst pools, (3) low-boreal bogs, (4) bogs and peat plateaus without thermokarst pools, (5) low-boreal dry poor fens, (6) wet moderate-rich fens, and (7) wet extremely rich fens. Thermokarst pools are features of a permafrost landscape where permafrost thawing and subsequent ice melting creates an uneven topography of mounds, sinkholes, caverns, and lake basins.

Locky et al. (2005) investigated black spruce (*Picea mariana*) swamps, fens, and bogs in the southern boreal region of Manitoba, Canada. They emphasized the distinction between black spruce swamps and other peatlands in the region pointing out their tendency to occur on gradual slopes, adjacent to water bodies, contain larger trees with significant cover, and to occur on shallower peat than other peatlands. Chemically, these swamps are similar to moderate-rich fens.

In another study that attempted to relate vegetation directly to water chemistry, Gorham and Janssens (1992) investigated two families of mosses (Sphagnaceae and Amblystegiaceae) at 440 sites across northern North America (see Figure 4-12). They found a clear bimodal split in their occurrence, with Sphagnaceae most often in low-pH peatlands (mode pH, 4.0–4.25) and Amblystegiaceae in high-pH peatlands (mode pH, 6.76–7.0).

Black Spruce Peatlands

One of the dominant forested wetlands in the world is the black spruce peatland of the taiga of Canada and Alaska. Black spruce (*Picea mariana*), often growing in association with tamarack (*Larix laricina*), is the tree species most associated with forested peatlands in the boreal regions of North America. These wetlands are estimated to encompass about half of the palustrine shrub–scrub wetlands in Alaska and cover an estimated 14 million ha in the state. Black spruce is mostly associated with ombrotrophic (bog) rather than minerotrophic (fen) communities. In bogs, it is found in associations with leatherleaf (*Chamaedaphne calyculata*), Labrador tea (*Ledum* spp.), laurel (*Kalmia latifolia*), blueberry (*Vaccinium* spp.), and bog rosemary (*Andromeda polifolia*). *Sphagnum* spp., of course, is found as ground cover in these bogs. In Alaska, common associations include *P. mariana* with *Vaccinium uliginonsum*, *Ledum groenlandicum*, and feathermoss (*Pleurozium schreberi*) and *P. mariana* with *Sphagnum* spp. and *Cladina* spp. (Post, 1996). In regions where permafrost is prevalent, black spruce wetlands often occur in paalsa hummocks.

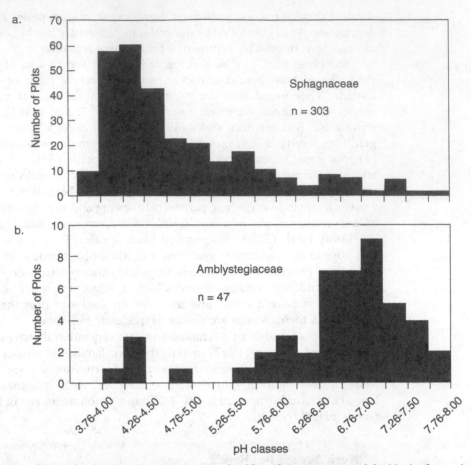

Figure 4-12 Distribution of two bryophyte families (Sphagnaceae and Amblystegiaceae) versus surface water pH for 440 peatland plots across North America. Plots were counted if they had at least one species of the family covering more than 25 percent of the total area. The bimodal pattern suggests a classification of peatlands based on moss vegetation. (After Gorham and Janssens, 1992)

Carolina Pocosins

In contrast to the more northern peatlands, the woody vegetation of pocosins found mostly near the southeastern USA coastal region of North and South Carolina is dominated by evergreen trees and shrubs. Two broad community classes have been identified, and their presence was related to fire frequency, soil type, and hydroperiod. A *Pinus–Ericalean* (pine and heath shrub) community develops on deep organic soils with long hydroperiods and frequent fire. Three associations within this community are (1) pond pine

(*continued*)

(*Pinus serotina*) canopy with titi (*Cyrilla racemiflora*) and zenobia (*Zenobia pulverulenta*) shrubs, (2) pond pine and loblolly bay (*Gordonia lasianthus*) canopy with fetterbush (*Lyonia lucida*), and (3) pond pine canopy with titi and fetterbush shrubs.

A *conifer–hardwood* community type is found on shallow organic soils with slightly shorter hydroperiods. Two associations in this group are (1) pond pine canopy with titi, fetterbush, red maple (*Acer rubrum*), and black gum (*Nyssa sylvatica*) shrubs; and (2) pond pine and pond cypress (*Taxodium distictium* var. *nutans*) canopy with red maple, titi, fetterbush, and black gum shrubs.

Peat buildup in pocosins has a major effect on hydrology and nutrients. Fire is a recurring influence that has been proposed as a major control of plant succession. Pocosins are found on the deepest peats that are always saturated at depth. Roots do not penetrate these deep peats into the underlying mineral soils. As a result, nutrients are limited and growth is stunted. It has been shown that the deepest peat often occurs at a topographic high point and, as pocosin elevations gradually decline, peat depths decline and pocosins become more minerotrophic (see Figure 4-13). This shift corresponds with structural changes from a shrubby "short pocosin" to more forested "tall pocosins" (Richardson, 2003).

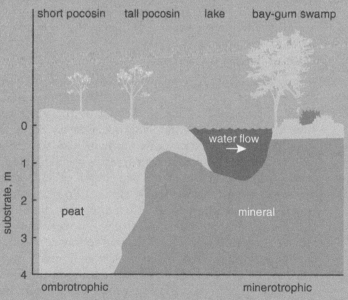

Figure 4-13   Cross-section of pocosin landscape showing short pocosin, tall pocosin, and bay-gum swamp in North Carolina. (*After Richardson, 2003*)

Peatland Adaptations

The vegetation in bogs and peatlands both controls and is controlled by its physical and chemical environment. Some of the conditions for which adaptations are necessary in peatlands are discussed here.

Waterlogging

Many bog plants, in common with wetland vegetation in general, have anatomical and morphological adaptation to waterlogged anaerobic environments. These include (1) the development of large intercellular spaces (aerenchyma or lacunae) for oxygen supply, (2) reduced oxygen consumption, and (3) oxygen leakage from the roots to produce a locally aerobic root environment. *Sphagnum*, conversely, is morphologically adapted to maintain waterlogging. The compact growth habit, overlapping leaves, and rolled branch leaves form a wick that draws up water and holds it by capillarity. These adaptations enable *Sphagnum* to hold water up to 15 to 23 times its dry weight.

Acidification of the External Interstitial Water

Sphagnum has the unique ability to acidify its environment, probably through the production of organic acids, especially polygalacturonic acids located on the cell walls (Clymo and Hayward, 1982). The galacturonic acid residues in the cell walls increase the cation exchange capacity to double that of other bryophytes. The adaptive significance of this peculiarity of *Sphagnum* is unclear. The acid environment retards bacterial action and, hence, decomposition, enabling peat accumulation despite low primary production rates. It has been suggested that the high cation exchange capacity also enables the plant to maintain a higher and more stable pH and cation concentration in the living cells than in the surrounding water.

Nutrient Deficiency Adaptations

Many bog plants have adaptations to the low nutrient supply that enable them to conserve and accumulate nutrients. Adaptations seen in bog plants include evergreenness; sclerophylly, or the thickening of the plant epidermis to minimize grazing; uptake of amino acids; and high root biomass. Some bog plants, notably cotton grass (*Eriophorum* spp.), translocate nutrients back to perennating organs prior to litterfall in the autumn. These nutrient reserves are available for the following year's growth and seedling establishment. The roots of other bog plants penetrate deep into peat zones to bring nutrients to the surface. Bog litter has been demonstrated to release potassium and phosphorus, often the most limiting nutrients, more rapidly than other nutrients, an adaptation that keeps these nutrients in the upper layers of peat. Many ericaceous plants have adapted to low concentrations of nitrogen by effectively utilizing ammonium nitrogen in place of limited nitrate nitrogen under low-pH conditions, by efficiently using nitrogen and even by utilizing organic nitrogen sources. Some bog plants also carry out symbiotic nitrogen fixation. The bog myrtle (*Myrica gale*) and the alder develop root nodules characteristic of nitrogen fixers and have been shown to fix atmospheric nitrogen in bog environments.

Carnivorous Plants

Another well-known adaptation to nutrient deficiency in bogs is the ability of carnivorous plants to trap and digest insects. This special feature is seen in several unique insectivorous bog plants, including the pitcher plant (*Sarracenia purpurea*; Figure 4-14) and sundew (*Drosera* spp.). A nutrient limitation study developed for *Sarracenia* in Minnesota showed, that although nutrient and insect additions did not increase biomass, there were respective nutrient increases in the leaves of the plant. A nutrient budget was developed and it was estimated that insect capture accounts for approximately 10 percent of the plant's nitrogen and phosphorus needs (Chapin and Pastor, 1995). Pitcher plants (*Sarracenia* spp.) are obligate host to more invertebrate species than any other bog plant (Rymal and Folkerts, 1982). A mosquito, a midge, two sarcophagid flies, and a mite are found in the water-filled pitcher in the plant, as shown in Figure 4-14. An aphid and three moths feed exclusively on the tissue. Other insects are associated with other parts of the plant.

Overgrowth by Peat Mosses

Many flowering plants are faced with the additional problem of being overgrown by peat mosses as the mosses grow in depth and in area covered. Adapting plants must

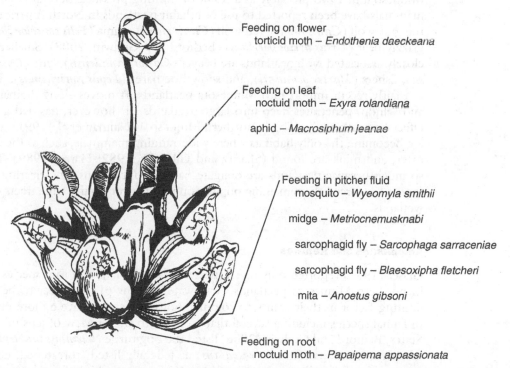

Feeding on flower
torticid moth – *Endothenia daeckeana*

Feeding on leaf
noctuid moth – *Exyra rolandiana*

aphid – *Macrosiphum jeanae*

Feeding in pitcher fluid
mosquito – *Wyeomyla smithii*

midge – *Metriocnemusknabi*

sarcophagid fly – *Sarcophaga sarraceniae*

sarcophagid fly – *Blaesoxipha fletcheri*

mita – *Anoetus gibsoni*

Feeding on root
noctuid moth – *Papaipema appassionata*

Figure 4-14 The pitcher plant (*Sarracenia purpurea*) including invertebrates that associate with the plant. (*After Damman and French, 1987*)

raise their shoot bases by elongating their rhizomes or by developing adventitious roots. Trees such as pine, birch, and spruce are often severely stunted because of the moss growth and poor substrate; they grow better on bogs where the vertical growth of moss has stopped.

Consumers

Mammals

The animal population in bogs is generally low because of the low productivity and the unpalatability of bog vegetation. Animal density is closely related to the structural diversity of the peatland vegetation. For example, forested peatlands tend to support the greatest number of small-mammal species, especially close to upland habitats. Large mammals tend to roam over larger landscapes and are, thus, not specific to individual peatland types. In northern Minnesota and New England, moose (*Alces alces*) are frequently found in small peatlands. White-tailed deer (*Odocoileus virginianus*) browse heavily in white cedar bogs in winter. Black bear (*Ursus americanus*) use peatlands for escape cover and for food. The woodland caribou (*Rangifer tarandus*) was the largest mammal that was largely restricted to peatlands, but it disappeared from Minnesota in 1936 probably as a result of hunting pressure. Several large predatory mammals have been reported to use or inhabit peatlands in North America including the gray wolf (*Canis lupus*), red wolf (*Canis rufus*), puma (*Felis concolor cougar*), and grizzly bear (*Ursus arctos horriblis*) (Bedford and Godwin, 2003). Smaller mammals closely associated with peatlands are beaver (*Castor canadensis*), lynx (*Lynx canadensis*), fishers (*Martes pennanti*), and snowshoe hares (*Lepus americanus*). The beaver is a fairly recent import into Minnesota peatlands. It moves along drainage ditches, and seldom penetrates deep into large peatlands. It, however, has had a significant effect on peatland flooding in northern Minnesota (Naiman et al., 1991). Wet forests are becoming the only habitats where wide-ranging mammals, such as the black bear, otter, and mink are found (Sharitz and Gibbons, 1982; Harris, 1989). This is not so much because peatlands are obligate habitats but because the clearing of upland forests has forced the remaining population into the remaining large tracts of forested wetlands.

Amphibians and Reptiles

Glaser (1987) reported only seven species of amphibians and four species of reptiles in northern Minnesota peatlands. Acid waters below pH 5 appear to be the major limiting factor in their ability to colonize bogs. Fens may have a more diverse array of faunal species including several that are rare. In their review of fens of the United States, Bedford and Godwin (2003) listed the bog turtle (*Clemmys muhlenbergii*) and eastern massasauga (*Sisturus catenatus*) as federally listed (threatened, endangered, or considered for listing) reptiles that use fens with high frequency. Other rare or uncommon species associated with fens include mole salamanders (*Ambystoma talpoideum*) and four-toed salamanders (*Hemidactylium scutatum*). These species frequent small mountain fens in the Appalachians (Murdock, 1994).

Birds

Many bird species are seen in peatlands during different times of the year. For example, Warner and Wells (1980) reported 70 species during the breeding season. Many of these are also common on upland sites, but a few depend on peatlands for survival including the Sandhill Crane (*Grus candensis*), Great Gray Owl (*Strix nebulosa*), Short-eared Owl (*Asio flammeus*), Sora (*Porzana carolina*), and Sharp-tailed Sparrow (*Ammospiza caudacuta*). In New England, as one moves from the Canadian border south, the species change, but the new species have analogous positions along the gradient (see Figure 4-15). Pocosins in the Southeast United States that support mature pond pine (*Pinus serotina*) can be inhabited by the endangered red-cockaded woodpecker (*Picoides borealis*) (Richardson, 2003).

Ecosystem Function

The dynamics of peatlands reflect the realities of the harsh physical environment and the scarcity of mineral nutrients. These conditions result in several major features:

- Bogs are systems of low primary productivity; fens are generally more productive; *Sphagnum* mosses often dominate bogs and other vegetation is stunted in growth.
- Bogs and fens are peat producers whose rates of accumulation are controlled by a combination of complex hydrologic, chemical, and topographic factors. This peat contains a great store of nutrients, most of it below the rooting zone and, thus, unavailable to plants.
- Low-nutrient peatlands in cold climates have developed several unique pathways to obtain, conserve, and recycle nutrients. The amount of nutrients in living biomass is small. Cycling is slow because of the low temperatures, the nutrient deficiency of the litter, and the waterlogging of the substrate. It is more active when peat production stagnates and when bogs receive increased nutrient inputs.

Primary Productivity

Major organic inputs to bog systems come from the primary production of the vascular plants, liverworts, mosses, and lichens. Among vascular plants, ericaceous shrubs and sedges are the most important primary producers, and much of this production is below ground. Mosses, especially sphagnum, account for one-third to one-half of the total production. Bogs and fens are usually less productive than most other wetland types and are generally less productive than the climax terrestrial ecosystems in their region, about half that of a coniferous forest and a little more than a third that of a deciduous forest (see Table 4-3). According to Pjavchenko (1982), forested peatlands

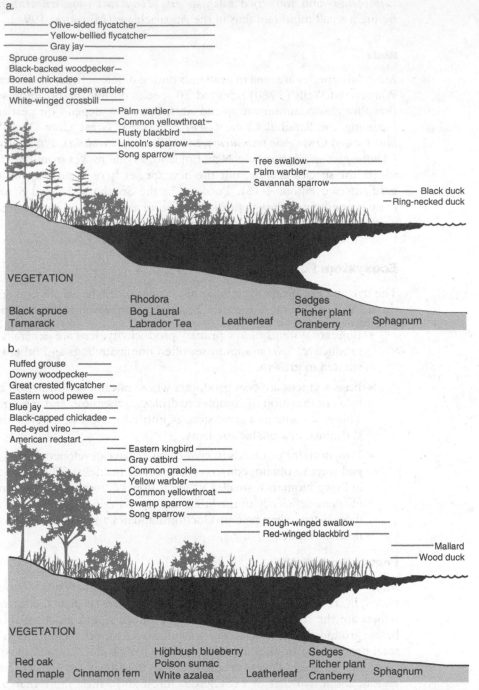

a.

Olive-sided flycatcher
Yellow-bellied flycatcher
Gray jay
Spruce grouse
Black-backed woodpecker
Boreal chickadee
Black-throated green warbler
White-winged crossbill
Palm warbler
Common yellowthroat
Rusty blackbird
Lincoln's sparrow
Song sparrow
Tree swallow
Palm warbler
Savannah sparrow
Black duck
Ring-necked duck

VEGETATION

Black spruce
Tamarack

Rhodora
Bog Laural
Labrador Tea

Leatherleaf

Sedges
Pitcher plant
Cranberry

Sphagnum

b.

Ruffed grouse
Downy woodpecker
Great crested flycatcher
Eastern wood pewee
Blue jay
Black-capped chickadee
Red-eyed vireo
American redstart
Eastern kingbird
Gray catbird
Common grackle
Yellow warbler
Common yellowthroat
Swamp sparrow
Song sparrow
Rough-winged swallow
Red-winged blackbird
Mallard
Wood duck

VEGETATION

Red oak
Red maple   Cinnamon fern

Highbush blueberry
Poison sumac
White azalea

Leatherleaf

Sedges
Pitcher plant
Cranberry

Sphagnum

Figure 4-15   Comparison of bird distribution, typical of a lake-border bog in the a. northern and b. southern parts of the northeastern United States. (*After Damman and French, 1987*)

Table 4-3 Net Primary Productivity of Peatlands in Europe and North America

Location	Type of Peatland	Living Biomass (g dry wt m^{-2})	Net Primary Productivity (g dry wt m^{-2} yr^{-1})	Reference
EUROPE				
Western Europe	general nonwooded raised bog	1,200	400-500	Malmer (1975)
Western Europe	forested raised bog	3,700	340	Moore and Bellamy (1974)
Russia	eutrophic forested bog	9,700–11,000	400	Pjavchenko (1982)
	mesotrophic forested bog	4,500–8,900	350	
	oligotrophic forested bog	2,200–3,600	260	
Russia	mesotrophic *Pinus-Sphagnum* bog	8,500	393	Bazilevich and Tishkov (1982)
England	blanket bog	659±53[a]		Forrest and Smith (1975)
England	blanket bog		635	Heal et al. (1975)
Ireland	blanket bog		316	Doyle (1973)
NORTH AMERICA				
Michigan	rich fen		341[b]	Richardson et al. (1976)
Minnesota	forested peatland	15,941	1,014	Reiners (1972)
	fen forest	9,808	651[b]	
Manitoba	peatland bog		1,943	Reader and Stewart (1972)
Alberta	bog		280[b]	Szumigalski & Bayley (1996a)
	poor fen		310[b]	
	moderately rich fen		360[b]	
	lacustrine sedge fen		214[b]	
	extremely rich fen		245[b]	
Alberta	bog		390[b]	Thormann & Bayley (1997)
	floating sedge fen		356[b]	
	lacustrine sedge fen		277[b]	
	riverine sedge fen		409[b]	
Quebec	poor fen		114[b]	Bartsch & Moore (1985)
	rich fen		335[b]	
	transitional fen		176[b]	

[a]Mean ± standard deviation for seven sites.
[b]Aboveground only.

produce a range of 260 to 400 g organic matter m^{-2} yr^{-1}, with the low value that of an ombrotrophic bog and the high value that of a minerotrophic fen. Malmer (1975) cited a typical range of 400 to 500 g m^{-2} yr^{-1} for nonforested, raised (ombrotrophic) bogs in western Europe. In contrast, Lieth (1975) estimated the net primary productivity in the boreal forest to average 500 g m^{-2} yr^{-1} and that in the temperate forest to average 1,000 g m^{-2} yr^{-1}. The estimate for boreal forests probably includes bog forests as well as upland forests. Annual aboveground productivity for control plots in a northern Minnesota fen ranged from 87 ± 2 to 459 ± 34 g m^{-2} yr^{-1} over a four-year period with belowground biomass (estimated at the last year) at 470 ± 79 g m^{-2} (Weltzin et al. 2005).

Comparative Study of Peatland Productivity in Canada

Thormann and Bayley (1997) investigated the aboveground net primary productivity of bogs, fens, and marshes in the southern boreal region of western Canada using similar field techniques (see Figure 4-16). They found that, when all aboveground strata were combined, the productivity of the bog, which was dominated by the moss *Sphagnum fuscum*, was 390 g m^{-2} yr^{-1}, about the

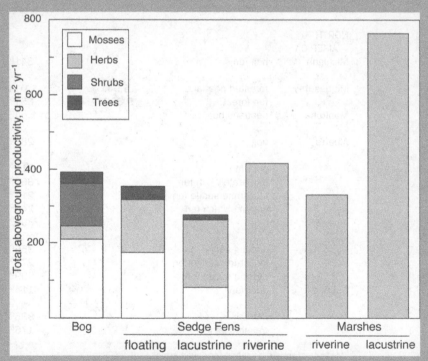

Figure 4-16 Net aboveground primary productivity in four peatlands and two marshes in southern boreal Alberta, Canada. (*After Thormann and Bayley, 1997*)

(continued)

same as the productivity measured in three fen sites (277–409 g m^{-2} yr^{-1}) and a riverine marsh (323 g m^{-2} yr^{-1}), but considerably lower than productivity in a lacustrine freshwater marsh (757 g m^{-2} yr^{-1}). Particularly for the fens and marshes, the absence of belowground productivity measurements probably greatly underestimated the overall productivity of these systems much more than that of the bog dominated by mosses.

The measurement of the growth or primary productivity of *Sphagnum* mosses presents special problems not encountered in productivity measurements of other plants. The upper stems of the plant elongate, and the lower portions gradually die off, become litter, and eventually form peat. It is difficult to measure the sloughing off of dead material to litter. It is equally hard to measure the biomass of the plant at any one time because it is difficult to separate the living and dead material of the peat. The following two methods for measuring *Sphagnum* growth give comparable results: (1) the use of "innate" time markers such as certain anatomical or morphological features of the moss and (2) the direct measurement of changes in weight. Growth rates for *Sphagnum* determined by these two techniques generally fall in the range of 300 to 800 g m^{-2} yr^{-1} (see Table 4-4). Although Damman (1979) and Wieder and Lang (1983) suggested that annual production should increase with decreasing latitude, only *S. magellanicum* shows such a trend in Table 4-4. Evidently, local and regional factors are more important than latitude.

It is generally expected that peatlands are nutrient limited. However, this can vary by plant species and community. Chapin et al. (2004) experimentally loaded N, P, and calcium carbonate (to raise pH) into a bog and fen in northern Minnesota and examined plant community and species productivity in response. In the bog, calcium carbonate and low additions of N (2 g N m^{-2} yr^{-1}) both increased aboveground net primary productivity (ANPP), while higher loads of N (6 g N m^{-2} yr^{-1}) actually inhibited growth. Fen graminoid growth responded to increased P additions. Within both wetlands, there were variable responses to experimental conditions among plant types and species. They surmised that although nutrient availability is low in peatlands this does not necessarily mean that they are nutrient limited.

Decomposition

The accumulation of peat in bogs is determined by the production of litter (from primary production) and the destruction of organic matter (decomposition). As with primary production, the rate of decomposition in peat bogs is generally low because of (1) waterlogged conditions, (2) low temperatures, and (3) acid conditions. In fact, the accumulation of peat in peatlands is due more to slow decomposition processes than to net community productivity. Besides leading to peat accumulation, slow decomposition leads to slower nutrient recycling in an already nutrient-limited system.

Table 4-4 Comparison of Selected Data on Production of *Sphagnum* Species in Order of Decreasing Latitude

Species[a]	Growth (mm/yr)	Production ($g\,m^{-2}\,yr^{-1}$)	Latitude (N)	Location	Mean Annual Precipitation (mm)	Mean Annual Temperature (°C)	Source
fus	1.4–3.2	70	68°22'	N Sweden	600	2.9	Rosswall and Heal (1975)
fus	—	250	63°09'	S Finland	532	3.5	Silvola and Hanski (1979)
fus	—	220–290	63°09'	S Finland	532	3.5	K. Tolonen, in Rochefort et al. (1990)
fus	7–16	195	60°62'	S Finland	632	4–4.8	Pakarinen (1978)
mag	9.5	70	59°50'	S Norway	1,250	5.9	Pedersen (1975)
ang	14.7	500	—	—	—	—	
fus	9.8	90	56°05'	S Sweden	800	7.9	Damman (1978)
mag	7.8	100	—	—	—	—	
mag	10–18	50–100	55°09'	England	1,270	9.3	S. B. Chapman (1965)
ang	28–34	110–240	54°46'	England	1,980	7.4	Clymo and Reddaway (1971)
mag	14–15	230	54°46'	England	1,980	7.4	Forrest and Smith (1975)
ang	—	240–330	—	—	—	—	
ang	38–43	110–440	54°46'	England	1,980	7.4	Clymo (1970)
fus	6–7	75–83	54°43'	Quebec	791	4.9	Bartsch and Moore (1985)
ang	4–17	19–127	—	—	—	—	T. R. Moore (1989)
fus	—	270	54°28'	England	1,375	7.4	Bellamy and Rieley (1967)
fus	30	424–801	54°20'	N Germany	714	8.4	Overbeck and Happach (1957)
mag	35–51	252–794	—	—	—	—	
ang	120–160	488–1,656	—	—	—	—	Reader and Stewart (1971)
fus	—	50	49°53'	S Manitoba	517	2.5	Pakarinen and Gorham (1983)
fus	17–24	240	49°52'	NE Ontario	858	0.8	Rochefort et al. (1990)
fus	7–31	69–303	49°40'	NW Ontario	714	2.6	
mag	11–34	52–240	—	—	—	—	
ang	20–39	97–198	—	—	—	—	
mag	62	540	39°07'	West Virginia	1,330	7.9	Wieder and Lang (1983)

[a] fus, *Sphagnum fuscum*; mag, *S. magellanicum*; ang, *S. angustifolium*.
Source: Adapted from Rochefort et al. (1990).

The pattern of *Sphagnum* decomposition is highest near the surface, where aerobic conditions exist. By 20 cm depth, the rate is about one-fifth of that at the surface, which is caused by anaerobic conditions. The bulk of the organic decomposition that does occur in peat bogs is by microorganisms, although the total numbers of bacteria in these wetland soils are much fewer than in aerated soils. As pH decreases, the fungal component of the decomposer food web becomes more important relative to bacterial populations. Verhoeven et al. (1994) used a cotton-strip decomposition method and found substantially lower decay rates in ombrogenous bogs compared to other peatlands and mineral soil wetlands. Total phosphorus (positive correlation) and soil organic matter (negative correlation) explained 75 percent of the decay rates. Thus, low nutrients and high organic matter (which keeps the soils reduced) contribute significantly to the low decay rates in the bogs. Szumigalski and Bayley (1996b) found the following progression of rates of decay of litter from peatlands in central Alberta: *Carex* > *Betula* > mosses. The highest decomposition rates were with plant material with the highest nitrogen content. Using a standard litter material of *Carex lasiocarpa*, litter losses were in the following order:

poor fen > wooded rich fen > bog > open rich fen > sedge fen

In the same region, Bayley and Mewhort (2004) compared peat-accumulating marshes and moderate-rich fens. Although these wetlands can appear similar, the decomposition rates for fens were notably slower than marshes which were attributed to the higher water levels in the marshes.

There has been considerable speculation about factors that give rise to patterned peatlands. The pattern of strings and flarks or hummocks and hollows, for example, appears to be related to differential rates of peat accumulation. Rochefort et al. (1990) determined that differential accumulation in a poor-fen system in northwestern Ontario, Canada, was caused more by differences in peat decomposition rates than by differences in primary production rates. They found that, even though the production rates of *Sphagnum* in hummocks were generally about equal to the rates in hollows or even lower than the rates in minerotrophic hollows, hummock species had slower decomposition rates than those of hollow species. As a result, peat accumulated faster on hummocks than in hollows, and hummocks may be expanding at the expense of hollows.

Peat Accumulation

The vertical accumulation rate of peat in bogs and fens is generally thought to be between 20 and 80 cm/1,000 yr in European bogs (Moore and Bellamy, 1974), although Cameron (1970) gave a range of 100 to 200 cm/1,000 yr for North American bogs and Nichols (1983) reported an accumulation rate for peat of 150 to 200 cm/1,000 yr in warm, highly productive sites. Malmer (1975) described a vertical growth rate of 50 to 100 cm/1,000 yr as typical for western Europe. Assuming an average density of peat of 50 mg/mL, this rate is equivalent to a peat accumulation rate of 25 to 50 g m^{-2} yr^{-1}. Hemond (1980) estimated a rapid accumulation rate of 430 cm/1,000 yr, eqivalent to 180 g m^{-2} yr^{-1}, for Thoreau's Bog, Massachusetts.

Bog Energy Flow Comparison

One of the earliest energy budgets for any ecosystem was determined in the classic study by Lindeman (1942) of Cedar Bog Lake, a small bog in northern Minnesota (see Figure 4-17a). Although this energy budget is crude, the main

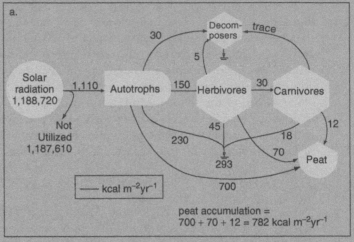

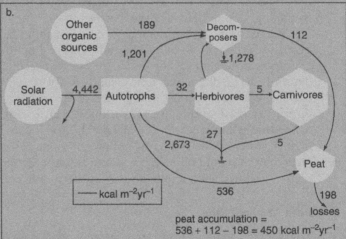

Figure 4-17 Diagrams of the energy flow in peatlands: a. Cedar Bog Lake, Minnesota, and b. a Russian transition peatland. Flow in kcal m⁻² yr⁻¹. Flows in (a) were originally published in calories whereas flows in (b) were published in gram dry weight and converted to energy as 4.5 kcal g⁻¹. (*a. After Lindeman, 1942; b. After Bazilevich and Tishkov, 1982; Alexandrov et. al., 1994*)

(*continued*)

features have stood the test of time. Very little of the incoming radiation (< 0.1 percent) is captured in photosynthesis. The two largest flows of organic energy are to respiration (26 percent) and to storage as peat (70 percent). Energy flow in the simplified food web is primarily to herbivores (13 percent), and about 3.5 percent goes to decomposers. As the following two more recent budget measurements show, the peat storage term is exceedingly high, and decomposition losses are probably underestimated.

Bazilevich and Tishkov (1982) and Alexandrov et al. (1994) presented a detailed energy flow through a mesotrophic (transition) bog in the European region of Russia (see Figure 4-17b). The bog is a sphagnum–pine (*Sphagnum girgenoshnii–Pinus sylvestris*) community containing shrubs such as bilberry (*Vaccinium myrtillus*). The total energy stored in the bog was estimated to be in excess of 137 kg dry organic matter/m^2, with dead organic matter (to a depth of 0.6 m of peat) accounting for 94 percent of the storage. Living biomass was 8.5 kg/m^2, or about 6 percent of the organic storage. Gross primary productivity was 987 g m^{-2} yr^{-1}, or about 4,400 kcal m^{-2} yr^{-1} (assuming 1 g organic matter = 4.5 kcal), with about 60 percent consumed by plant respiration. The distribution of the net primary production came from trees (39 percent), algae (28 percent), shrubs (21 percent), mosses and lichens (9 percent), and grasses (3 percent). The net primary production was primarily consumed by decomposers; much less was consumed in grazing food webs. Net accumulation of peat was 100 g m^{-2} yr^{-1} (or about 450 kcal m^{-2} yr^{-1}). Losses other than biotic decomposition, which accounted for most of the loss of organic matter, were chemical oxidation and surface and subsurface flows.

Comparison of these two energy budgets from Russia and the United States, carried out several decades apart, illustrates several points. First, Lindeman's Cedar Bog was approximately one-fourth as productive as Bazilevich and Tishkov's Russian peatland, a possibility, given that the Russian site was described as transitional between a bog and a fen. Second, the American bog accumulated more peat than did the Russian peatland. These results could reflect either the sophistication of the measuring techniques for the time or what happens to peatlands as they transition from fens to true bogs. Lindeman shows a high percent of the productivity stored permanently as peat. Assuming 50 g/L as the density of peat, Lindeman's peat accumulation results in a high rate 350 cm/1,000 yr, whereas the Russian study has a more reasonable 200 cm/1,000 yr. North American and European studies generally bracket a range of 50 to 200 cm/1,000 yr (see the preceding section "Peat Accumulation").

Nutrient Budgets

Nitrogen

Nitrogen budgets for two peatlands—Thoreau's Bog in Massachusetts and a perched raised-bog complex in Minnesota—make an interesting comparison (see Figure 4-18). Although total nitrogen input is comparable in both systems, the Minnesota perched-bog system catches some runoff from surrounding uplands, whereas nitrogen fixation is the largest source of biologically active nitrogen in Thoreau's Bog.

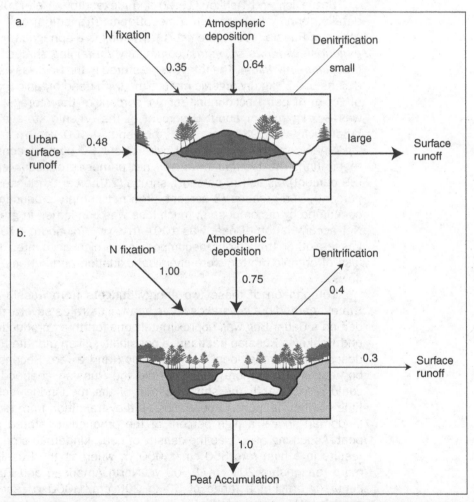

Figure 4-18   Nitrogen budgets for two northern ombrotrophic bogs: a. perched raised-bog complex in northern Minnesota and b. small floating-mat sphagnum bog (Thoreau's Bog) in Massachusetts. Values are in g N m^{-2} yr^{-1}. (*Data from Urban and Eisenreich, 1988; Hemond, 1983*)

Otherwise, the budgets are remarkably similar despite the differences in bog type and location. Both accumulate nitrogen in peat and lose a significant portion through runoff. Denitrification is an uncertain term.

Peatlands have often been identified as diverse wetlands. Increased nutrient loading appears to reduce that diversity. Drexler and Bedford (2002) traced various nutrients in a small peatland adjacent to a farm field in central New York. They found that the farm field was a nutrient source with P and K loading coming primarily from overland flow while N loading was primarily delivered via groundwater. Peat concentrations of P and K, along with groundwater flux of NO_3-N and NH_4-N, were all negatively correlated with plant diversity. Nutrient loading promoted the growth of taller, monotypic stands of vascular plants (*Calamagrostis canadensis*, *Carex lacustris*, *Epilobium hirsutum*, and *Typha latifolia*) at the expense of plant diversity.

Peatlands appear to have the capacity for denitrification (Hemond, 1983), but the magnitude *in vivo* is uncertain. Some recent studies have investigated nitrogen dynamics in peatlands. Wray and Bayley (2007) estimated annual denitrification rates for boreal marshes and fens in Alberta at 11 and 24 g N m^{-2} y^{-1}, respectively with N_2 being the dominant product.

Bridgham et al. (2001) examined nutrient availability along an ombrotrophic-minerotrophic gradient for 16 peatlands in northern Minnesota. They found that N availability generally increased along the gradient. They detected seasonal patterns associated with N availability with NO_3-N being more available in the summer and NH_4-N more available in the winter. Highest P availability was found in minerotrophic swamps and beaver meadows with the least availability found in fens and bogs.

Carbon

Carbon budgets for peatlands have drawn a great deal of interest, given the importance of these ecosystems in global carbon dynamics (see Mitsch and Gosselink, 2007, Chapter 10 for more discussion on this topic). High latitude peatlands are known to store tremendous amounts of carbon. In the Western Siberian lowlands, Kremenetski et al. (2003) estimated that peatlands have an average peat depth of 2.6 m and store greater than 53.8 million metric tons of carbon. It is generally accepted that boreal peatlands were once carbon sinks, but there is little consensus that they are contemporary sinks. Carbon budgets have been developed for small peatlands and for large peatland-dominated watersheds (Rivers et al., 1998). The latter, a 1,500-km^2 watershed in the Lake Agassiz peatlands in Minnesota, illustrated that the peat watershed had a net carbon storage of 12.7 g C m^{-2} yr^{-1} but that there was a tenuous balance between the watershed being a source and a sink of carbon. Inflows of carbon are groundwater, precipitation, and net community productivity, whereas outflows are groundwater and surface flow and outgassing of methane. It was estimated from a companion study (Glaser et al., 1997b) that peat is accumulating at a rate of 1 mm/yr (100 cm/1,000 yr). This budget illustrates the importance of accurate hydrologic measurements as well as biological productivity measurements in determining accurate nutrient budgets.

Studies have identified the importance of relative water levels to the carbon dynamics of peatlands. Bubier et al. (2003) examined CO_2 exchange among different wetland types (bog, fen, and beaver pond margin) in an expansive peatland in Ontario, Canada. During a dry summer, CO_2 uptake was diminished and most of the wetland became a net source of carbon. During a wetter summer, respiration rates were regulated less by water table position and more by peat temperature.

When predicting the response that peatlands will have to climate changes it is important to recognize that long-term conditions may not necessarily match the results of short-term studies. Weltzin et al. (2000) examined the influence of infrared loading and water levels by experimentally manipulating Minnesota peatland monoliths (see Chapter 5). Their results showed different responses for bogs and fens. As water levels rose in the bogs, bryophyte production increased and shrub production decreased. However, higher infrared radiation promoted greater shrub growth and decreased graminoid production. Fens showed enhanced graminoid production under higher infrared loading. Their results suggest that a changing climate may cause shifts in peatland plant composition and structure, and these shifts need to be considered when making predictions of long-term climate change impacts.

Lafleur et al. (2003) used eddy covariance measurements to examine net ecosystem CO_2 exchange (NEE) in an ombrotrophic bog near Ottawa, Canada. Over a three-year period, the bog provided relatively similar daytime uptake rates in the summer and functioned as an annual sink of CO_2 (~ 260 g CO_2 m^{-2} yr^{-1}). However, in a very dry year with lower water tables, that annual rate was only 34 g CO_2 m^{-2} yr^{-1}. The authors also noted the importance of winter CO_2 flux, which was estimated at 119–132 g CO_2 m^{-2} out of the wetland or 30–70% of the net CO_2 uptake during the previous growing season.

Methane emissions from peatlands have also been studied closely because of the potential influence as greenhouse gas and the expansive store of carbon in peatlands. Annual precipitation and water table position also seems to be a primary controller of methane flux in peatlands (Huttunen et al., 2003; Smemo and Yavitt, 2006). Methane emission from a Finnish minerotrophic peatland ranged between 8 and 330 mg m^{-2} d^{-1} and there was a positive correlation between methane emission and water table level (Huttunen et al., 2003). Bubier et al. (2005) found a similar seasonal average range of CH_4 emission (10-350 mg m^{-2} d^{-1}) but emphasized there was considerable spatial variability. They too found a significant relationship between water table position and mean CH_4 flux (see Figure 4-19). They noted the log linear relationship between CH_4 flux and water table position meaning that only a small increase in water table depth was needed to increase CH_4 emission substantially. In their review of the forested peatland literature, Trettin et al. (2006) concluded that decreased water table levels will result in decreased CH_4 emission and increased CO_2 emission from the peat surface. They emphasized this does not mean that these peatlands will necessarily decrease in their soil C-pool as the difference may be made up by changes in plant succession and increased productivity.

Fewer studies have looked at carbon emission in tropical peatlands. Jauhiainen et al. (2005) looked at the carbon flux in a mixed tropical forest peatland in Indonesia.

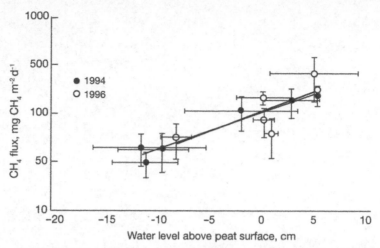

Figure 4-19 Methane flux as a function of mean water level in Canadian peatlands. Water table indicates depth below (negative) or above (positive) the peat surface. Error bars are standard deviations. (*After Bubier et al., 2005*)

The forest is comprised of undulating hummocks and hollows, with hummock surfaces 20–30 cm above the hollows and comprised of tree root masses that have become covered with debris. They found that CO_2 emission from the hummocks were relatively stable and emission from the hollows diminished as water tables rose. Carbon dioxide was the dominant form of carbon flux. They estimated annual emissions from the forest at 3493 ± 316 g-CO_2 m^{-2} y^{-1} and $< 1.36 \pm 0.57$ g-CH_4 m^{-2} y^{-1}. The ratio of hummock-hollow surface also influenced C flux as areas with higher hummock coverage contributed more CO_2 and less CH_4.

Recommended Readings

Strack, M., ed. 2008. Peatlands and Climate Change. International Peat Society, Jyvaskyla, Finland, 223 pp.

Wieder, R. K. and D. H. Vitt, eds. 2006. Boreal Peatlands Ecosystems. Springer-Verlag, Berlin, 435 pp.

Chapter 5

Ecosystem Approaches to Wetland Science

The study of wetlands as ecosystems is presented in this chapter through three distinct approaches—mesocosms, or small replicated ecosystems; whole-ecosystem studies that attempt to investigate entire ecosystems in all or most of their complexity, often in an experimental fashion; and mathematical modeling, by which wetlands are "experimented with" mathematically rather than in labs or in the field. Mesocosm studies have been conducted for salt marshes, freshwater marshes, and peatlands and even for mangrove swamps—although for this last option, mangrove seedlings were used and not full trees. There have been a notable number of published mesocosm studies of locations such as the Florida Everglades and the Olentangy River Wetland Research Park in Ohio. Mesocosm studies are small in scale, usually 1 to 3 m² for the individual mesocosms tubs or plots, replicated to allow better statistical analysis, and simple in their ecosystem structure. They usually have a one- to three-year life, after which they are limited by artifacts. Whole ecosystem studies of wetlands described here, dating from the 1960s through the present, include coastal ponds in North Carolina, cypress swamps in Florida, lacustrine freshwater marshes in Manitoba, and riverine wetlands in Illinois and Ohio. These studies had relatively large (> 0.5 ha) experimental units and were studied for a number of years by many researchers. Whole ecosystem studies need to be carefully managed because there is a tendency for the research team to study the parts but not the whole. In addition, these studies are often quite expensive in time and money to maintain. The third alternative, mathematical modeling, is in one sense the most flexible because it can be used to study large spatial scales, such as the entire Florida Everglades, and because the time frame can range from days to centuries. Modeling does require mathematical skills as well as a good understanding of ecology. While there are never sufficient data to have a perfectly calibrated and validated model, much is learned in the process of modeling, not just in the final simulations. Models have been developed for salt marshes, freshwater swamps, freshwater marshes, and peatlands. Many are elegantly simple in the mathematics and quite a few of the models presented here were simulated with software such as STELLA that was designed to model dynamic systems with relative ease.

Wetland ecosystems can be studied as systems on different scales—as microcosms, mesocosms, whole ecosystems, and landscapes. The wetland ecosystems that we presented in the previous three chapters have all been investigated at these scales. We have chosen to present three different system approaches for system's approaches to wetlands: mesocosms, whole-ecosystem studies, and mathematical models. Each approach has distinct advantages and limitations. Mesocosm studies are relatively easy to implement, can be completed by one or two graduate students, and can obtain meaningful results if the ecosystem is relatively simple and short-lived. Whole-ecosystem studies require teams of researchers, a commitment of several years, and major financial support. They are best managed at permanent research locations so that long-term monitoring can be part of the study. Mathematical models, if adequate data sets are available, can be one of the most cost-effective ways to investigate ecosystem behavior and to experiment in a mathematical way with an ecosystem that would otherwise be impossible with full-scale or mesocosms-scale studies.

Mesocosms

One of the most effective ways to "simulate" wetland ecosystems is with the use of mesocosms. "Meso-cosm" means "little world." Mesocosms are middle-size experimental units, too small to be considered wetlands but too large to fit into most laboratories. Essentially, they can be created to represent little simplified wetlands, and because they are small, they can be replicated for statistical power in interpreting the results. Eugene Odum (1984) presented an eloquent argument for doing ecological science with the aid of mesocosms, which he described as a "middle-sized world" that falls between reductionistic ecology that is often done in the laboratory and holistic systems science that is done in the complex real world of ecosystems. Even 25 years ago, he saw that science was already becoming quite reductionistic, a trend that continues to today, at the expense of understanding how systems, such as ecosystems, really work as a whole. His description of the "middle ground" of mesocosms studies is worth repeating:

> Mesocosms can be replicated, yet possess a degree of realism not possible in laboratory systems. Best of all, parts and wholes can be investigated simultaneously by a team of researchers. For example, senior investigators can focus on ecosystem-level behavior that requires longer time spans, while students carry out intensive shorter-term studies ... We might think of these outdoor ecosystems as living, working models. Like all models, they represent simplifications of real world systems. Accordingly, projections must be made with great caution.
> (Odum, 1984)

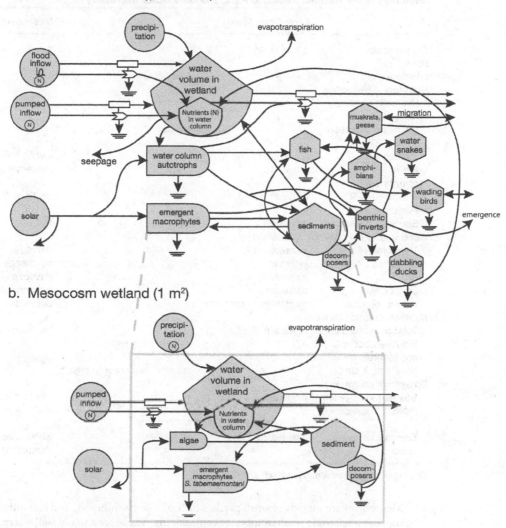

a. Full-scale wetland (10,000 m²)

b. Mesocosm wetland (1 m²)

Figure 5-1   Comparison of the structure and major energy and nutrient pathways of a. full-scale experimental freshwater marshes with b. an adjacent set of freshwater marsh mesocosms. Study done at Olentangy River Wetland Research Park at Ohio State University. (*After Ahn and Mitsch, 2002b*)

Wetland mesocosms, while simplifications of full-scale wetlands, offer the advantage of lower cost and replication not afforded by full-scale ecosystem studies. Figure 5-1 and Table 5-1 compare a full-scale wetland ecosystem ($\sim$10,000 m²) with an adjacent set of 1-m² wetland mesocosms in Ohio. The scale and complexity of these two systems were clearly different, yet many of the findings at the mesocosms scale mirrored those of the full scale. As we concluded from that study:

Table 5-1 Comparison of Full-Scale Freshwater Marsh and Mesocosm Marsh at the Olentangy River Wetland Research Park, The Ohio State University

	Large-Scale Marsh		Mesocosm Marsh
Spatial scale	$10,000 \text{ m}^2$		1 m^2
Source soil		identical	
Source water		identical	
Hydrology / Hydraulics			
Hydraulic Loading Rate (HLR)		similar	
Hydraulic Retention Time (HRT)		similar	
Turbulence	moderate to high		low
Mixing	moderate		moderate to high
Macrophytes			
Effective NAPP	$380 \text{ g m}^{-2} \text{ yr}^{-1}$		$353 \text{ g m}^{-2} \text{ yr}^{-1}$
Percent cover	$\sim 55\%$		$\sim 100\%$
Species richness	moderate		low
Water quality change through wetland			
Temperature	increase		decrease
Dissolved oxygen	increase		no change or decrease
pH	increase		increase
Conductivity	decrease		increase or no change
Redox potential	no change or decrease		decrease
Nutrient retention capacity			
Total phosphorus	moderate (fluctuating)		low (decreasing)
Soluble reactive phosphorus		similar (very high)	
Nitrate plus nitrite		similar (moderate to high)	
Ecosystem complexity			
Spatial heterogeneity	moderate		low to none
Biological complexity	moderate to high (developing)		low
Effective temporal scale	long (years)		short (weeks to months)

Source: Adapted from Ahn and Mitsch (2002b).

Mesocosms are models of small patches of the large counterpart, and can only support a relatively small number of components. These are usually soil, water, macrophytes, and microbes. Our mesocosms, by their very nature, could not contain fish, waterfowl, muskrats, amphibians, wading birds, or other mobile animals ... Wetland function is controlled not only by hydrology and nutrient inflows but also by biotic feedbacks from the biota (e.g. detrital buildup, transpiration, eatouts, sediment excavation, stream damming, etc.) ... Adey et al. (1996) argued, however, that functions of the system should not be judged on presence of particular components but rather on presence of major structural components allowing self-design and self-organization of the system which support a specific functioning (i.e., phosphorus retention) of the system studied. Our mesocosm wetlands contained the same relative forcing functions (sunlight, water and nutrient inflow) and main components (plants, sediments and water) as those in the large wetland that allowed self-organization of the system to manifest itself, thus simulating a large wetland for their macrophyte production and nutrient retention capacity over a relatively short period of time. In that sense, the mesocosms do offer a reasonable model of the large system in the temporal scale even without all of the complexity. (Ahn and Mitsch, 2002b)

Just as we divided wetland ecosystems in this book into coastal wetlands, inland marshes and swamps, and peatlands, we describe here some of the applications of mesocosms to understanding wetlands and how they work in similar categories. Perhaps because marsh-type wetlands, as shallow water or saturated soil systems, can be more easily reconstructed as mesocosms than can open oceans, lakes, or forests, there have been many applications of the mesocosms approach for understanding wetlands. Table 5-2 lists some of the many wetland mesocosms studies that have been published in the past 20 years and the questions that these studies were investigating. A brief summary of some of these studies is provided below.

Coastal Wetland Mesocosms

One of the fundamental problems in developing coastal wetland mesocosms is creating the day-to-day tidal cycle. This is usually done with a pump and timer arrangement. Once that plumbing problem is solved, the mesocosms are similar in operation to freshwater marsh systems. Several coastal wetland mesocosms have been used to investigate the interactions of ecosystem plants and animals on biogeochemical cycling. In one of the first mesocosm studies of this kind, Gribsholt and Kristensen (2002) investigated the effect of the polychaete worm (*Nereis diversicolor*) and the macrophyte *Spartina anglica* on salt marsh biogeochemistry in mesocosms. They summarized: "The presence of both flora and fauna produced the largest increase in oxic as well as total microbial mineralization rates. In conclusion, the mesocosm approach simulated natural salt marsh conditions well, and the results provide good evidence for the interactions between flora and fauna as well as their impact on sediment geochemistry." (Gribsholt and Kristensen, 2002)

Continuing in the same direction of biotic impact on biogeochemistry, Kristensen and Alongi (2006) describe a coastal mangrove mesocosm study on the effects of fiddler crabs (*Uca vocans*) and black mangroves (*Avicennia marina*) on carbon, sulfur, and iron biogeochemistry (see Figure 5-2). Their mesocosms were tubs 2.2 m x 0.7 m and 50 cm deep (see Figure 5-2a) with semi-diurnal tide simulated by seawater pumped from a 800 L reservoir with a time-controlled pump. Water was recycled from the reservoir and the reservoir was replenished every two weeks with clean seawater. Of the four mesocosms, two were planted with *Avicennia* saplings and two were left unplanted. Then two mesocosms (one planted, one not) had 96 fiddler crabs introduced in a 2 × 2 experiment (see Figure 5-2b). *Avicennia* saplings grew better in the presence of crabs but there were more microalgal mats instead of mangrove growth in the absence of crabs. Both crabs and *Avicennia* oxidized the surface sediments, increasing oxidized iron there. Sulfate reduction was more important in the presence of crabs, particularly when plants were present. The authors conclude

Table 5-2 Mesocosms Used to Model and Experiment with Wetland Ecosystems

Ecosystem simulated	Location	Purpose	References
Freshwater marsh	San Diego, California	To investigate the role of hydroperiods on marsh retention of nutrients and metals	Busnardo et al., 1992; Sinicrope et al., 1992
Freshwater and salt marsh	Louisiana	Shifts in vegetation with salinity and water level changes in coastal marshes	Baldwin and Mendelssohn, 1998
Peatlands	Minnesota	Peatland monolith mesocosms used to study role of peatlands in climate change	Bridgham et al., 1999; Weltzin et al., 2000; Noormets et al., 2004
Freshwater marsh	Olentangy River Wetland Research Park, Ohio	Importance of different genotypes of *Juncus effusus* from different regions	Weihe and Mitsch, 2000
Freshwater marsh	Olentangy River Wetland Research Park, Ohio	To compare competitive growth of two wetland macrophyte species in low and high nutrients	Svengsouk and Mitsch, 2001
Freshwater marsh	Olentangy River Wetland Research Park, Ohio	To investigate the use of sulfur scrubber material as liners for treatment wetlands	Ahn et al., 2001; Ahn and Mitsch, 2002a,b
Peatland	Waikato region, North Island, New Zealand	To determine the role of fertilizer, seed additions and cultivation techniques for peatland restoration	Schipper et al., 2002
Salt marsh	Kerteminde, Denmark	To estimate the effects of benthic fauna (*Nereis diversicolor*) and vegetation (*Spartina anglica*) on wetland biogeochemistry	Gribsholt and Kristensen, 2002
Freshwater stormwater marsh	Everglades, Florida	To determine the type of vegetation (emergent vs. submersed) and hydrology (continuous flooding vs. periodic drawdown) that is most effective in phosphorus removal	White et al., 2004, 2006

Table 5-2 *(continued)*

Ecosystem simulated	Location	Purpose	References
Freshwater marsh	Olentangy River Wetland Research Park, Ohio	To investigate effects of hydrologic pulsing on productivity and nutrient cycling	Anderson and Mitsch, 2005
Mangrove swamp	North Queensland, Australia	To estimate the effects of benthic fauna (*Uca vocans*) and mangroves (*Avicennia marina*) on wetland biogeochemistry	Kristensen and Alongi, 2006
Freshwater treatment marsh	Florida	To test addition of alum to wetlands to improve phosphorus retention	Malecki-Brown et al., 2007
Freshwater marsh	Ohio	To assess the importance of wetland plant functional diversity on belowground biomass, methane production, and denitrification potential	Bouchard et al., 2007
Mangrove swamp and salt marsh	National Wetlands, Research Center, Louisiana	To see the effect of CO_2 enrichment on competition between tropical mangroves and temperate salt marshes	McKee and Rooth, 2008
Freshwater marsh	Saitama, Japan	To investigate the effect of *Typha* and water level fluctuations on N and C removal in subsurface flow	Sasikala et al., 2008
River floodplains	Monoliths from Netherlands and Poland	To determine short-term effect of summer flooding on floodplain vegetation from modified and pristine floodplains	Antheunisse and Verhoeven, 2008
Freshwater marsh	Florida Everglades	To estimate the ecosystem fate of introduced nitrogen in Florida Everglades with insitu mesocosms	Wozniak et al., 2008
Freshwater marsh	Olentangy River Wetland Research Park, Ohio	To investigate effect of hydrology and hydric soils on methane emissions	Altor and Mitsch, 2008a

a.

b.

Figure 5-2 Illustrations from a coastal wetland mesocosm study (Kristensen and Alongi, 2006) that investigated the effects of a. black mangrove seedlings (*Avicennia marina*) and b. fiddler crabs (*Uca vocans*) on carbon, sulfur, and iron biogeochemistry of coastal mangrove swamps. (*Photos provided courtesy of E. Kristensen*)

that tidal mesocosms "are well suited for studying the influence of plants and animals on the biogeochemistry of intertidal sediments" and that the mesocosms provide a simple mimic of natural mangrove sediments. Mesocosm studies of this type could not be done with mature mangrove trees because of the number of years it would take for the trees to grow and the scale needed to accommodate the trees.

McKee and Rooth (2008) also developed coastal wetland mesocosms to investigate the effects of elevated CO_2 on the competition of tropical mangrove trees (*Avicennia germinans*) with temperate salt marsh grass (*Spartina alterniflora*). Their study was conducted in four 12-m² greenhouses that have appropriate temperature and CO_2 concentration controls (see Figure 5-3). Two greenhouses had elevated 720 ppm CO_2 and two had ambient (365 ppm). Within each greenhouse, there were 18 plots (6 each of *Spartina* monoculture, *Avicennia* monoculture, and *Spartina-Avicennia* mixtures). The mangroves were introduced by seed introduction, the *Spartina* by transplants from native marshes. The researchers postulated that increased CO_2 may favor the mangrove (a C_3 plant) over *Spartina* (a C_4 plant). By itself, *Avicennia* production increased by 35 percent in high CO_2 concentrations but in the presence of *Spartina*, its growth was severely curtailed. Combining these results with field tests, the authors conclude that mangroves will not supplant salt marshes due to CO_2 increases alone, but may do so if other climate changes result from the CO_2 increases. By the end of the experiment, the mangroves were 1 m tall.

Freshwater Marsh Mesocosms

Freshwater marshes provide the easiest application of mesocosm models for investigating wetland function. First, marsh plants can be grown to represent natural

Figure 5-3 Wetland Elevated CO₂ Experimental Facility at the USGS National Wetlands Research Center in Lafayette, Louisiana, which consists of four 12-m² greenhouses with appropriate temperature and CO₂ concentration controls for greenhouse gas experiments. McKee and Rooth (2008) conducted a coastal wetland mesocosm study on competition between mangroves and salt marsh grass in elevated CO₂ conditions in this facility. (*Photos provided courtesy of K. McKee*)

conditions in a relatively short time (often one growing season) as opposed to any woody vegetation in forested swamps that would take years to develop. Second, freshwater marshes do not have to be treated with salt water (it is difficult to maintain large reservoirs of salt water in salt marsh studies) or with daily tidal regimes as is the case in tidal salt marshes. Mesocosm studies done of freshwater marshes in two distinctly different climates—humid temperate (Ohio and Netherlands) and humid subtropical (Florida)—are discussed here.

Mesocosms at the Olentangy River Wetland Research Park, Ohio

Wetland functions have been investigated with wetland mesocosms at the Olentangy River Wetland Research Park at Ohio State University since the research park was started in the mid-1990s. The mesocosms used there (see Figure 5-4) are approximately 1 m^2 in area and are sunk into the ground

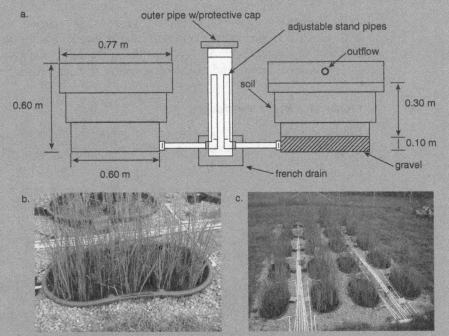

Figure 5-4 Freshwater marsh mesocosms established at the Olentangy River Wetland Research Park at The Ohio State University including a. sketch of paired mesocosms with outflow system, b. an individual mesocosms tub with macrophytes, and c. a set of 20 mesocosms with inflow system installed. These mesocosms have been used in experiments to compare macrophytes competition in high nutrient conditions (*Sveng-souk and Mitsch, 2001*), evaluate the effects of flue gas desulfurization (FGD) waste material for lining wastewater treatment wetlands (*Ahn et al., 2001; Ahn and Mitsch, 2001, 2002a,b*), and investigate the importance of hydrologic pulsing on plant productivity (*Anderson and Mitsch, 2005*) and methane emissions (*Altor and Mitsch, 2008a*).

(*continued*)

so that soil temperatures approximate natural conditions and to avoid winter root kill and overheated soils in the summer. A simple backflow system was designed that could both serve as a water level regulator and allow water to flow out of the mesocosms if flowthrough conditions are desired. Experiments can be run with any number of mesocosms, depending on the research topic. Four of the published experiments in these mesocosms sets at the Olentangy River Wetland Research Park are described here to illustrate the variety of wetland ecosystem topics that can be investigated with mesocosms.

Svensouk and Mitsch (2001) ran an experiment on the effects of nitrogen, phosphorus, and combined nitrogen and phosphorus on aboveground and belowground productivity of *Typha* sp. (cattails) and *Schoenoplectus tabernaemontani* (soft-stem bulrush) in both symmetric and asymmetric studies. These two plants had clearly been the most dominant species in the nearby experimental wetlands at the Olentangy River Wetland Research Park prior to the mesocosm study. In a symmetric competition study, *Typha* and *Schoenoplectus* were planted equally in each treatment. *Typha* responded in the first year to increased combination of nitrogen and phosphorus more than did *Schoenoplectus* in the symmetric study. Nutrient additions were relaxed in the second year and *Schoenoplectus* responded in both belowground and aboveground growth better than did *Typha*. The greater growth of *Typha* when nitrogen and phosphorus were applied in combination in the symmetric competition suggests that the aggressive invasion of this plant in many freshwater marshes throughout the Midwestern United States could be due to both nutrients in combination rather than only one.

Freshwater marsh mesocosms were used to simulate the potential use of FGD (flue gas desulphurization) waste product from sulfur scrubbers as a liner material in wetlands in Ohio and other coal-burning regions (Ahn et al., 2001; Ahn and Mitsch, 2001, 2002a,b). Liners are generally necessary to prevent water from seeping through the bottom of constructed wetlands that are created for improving water quality. The study showed not only that the material served as a good aquiclude to protect groundwater but also the calcium-dominated FGD material caused those wetlands with that material as liners to be more effective in phosphorus retention. It was during that same experiment that the control sites (non-lined) were compared to similar conditions in the adjacent full-scale, flowthrough marshes at the Olentangy River Wetland Research Park (Figure 5-1; Table 5-1; Ahn and Mitsch, 2002b). While the hydrology was similar when scaled per unit area, there were differences in water quality, particularly temperature and dissolved oxygen, at the two scales, and these differences, in turn, may have affected phosphorus.

(*continues*)

(continued)

Ecosystem complexity and spatial heterogeneity in the mesocosms were much lower than conditions found in the full-scale wetland, but they simulated the full-scale wetlands reasonably well.

Two more recent mesocosm studies investigated the importance of hydrologic pulsing on freshwater marsh function. Anderson and Mitsch (2005) investigated the effect of hydrologic pulsing on herbaceous productivity and nutrient uptake of marsh ecosystems. Half of the mesocosms were planted with *Typha angustifolia* (cattail) and the other half were planted with *Schoenoplectus tabernaemontani* (soft-stem bulrush). In each of those treatments, half of the wetlands were subjected to pulsing hydrology, while the others were given steady-flow conditions. While the cattail mesocosms were more productive than the bulrush mesocosms, there were no differences in productivity in the two hydrologic conditions in this study. The N:P ratios were higher in the vegetation in pulsing conditions, suggesting that the steady-flow wetlands were more limited by nitrogen. Altor and Mitsch (2008a) continued the theme of hydrologic pulsing by investigating how it affects methane emissions from freshwater marsh mesocosms. A second variable added to the experiment was soil type—hydric soil extracted from a nearby wetland was introduced to half of the mesocosms while the other half had upland soils as a starting point. Methane emissions were lower in mesocosms that had pulsing hydrology than in those with continuous inundation when hydric soil was present, a normal occurrence in natural marshes. Overall, methane emissions were higher in hydric soil mesocosms than in those that had non-hydric soils; this difference was attributed to the general absence of a robust community of methanogens in the non-hydric soils. This study, which used gas-sampling chambers on mesocosms, confirmed that wetland mesocosms can be effective in comparing greenhouse gas emissions from freshwater marshes, as had been found previously in northern peatlands (see below).

Netherlands River Floodplains

Antheunisse and Verhoeven (2008) investigated potential effects of returning floodplains in the Netherlands to a more natural hydrology with mesocosms. They wanted to see the effects of summer river flooding on plant diversity on monoliths from an impacted floodplain of the Rhine River in Germany and the more pristine Narew River from Poland. Their study simulated the flooding by simply submerging monoliths from the two regions in large circular containers (see Figure 5-5). They found that 14-day summer flooding decreased biomass in the impacted Rhine floodplain but it was unchanged in the Polish floodplain. Plant diversity in both systems decreased because of the flooding. They suggested that the return of a more natural hydrology that would include summer flooding would lead to a short-term reduction in productivity and other changes in species abundance and dominance.

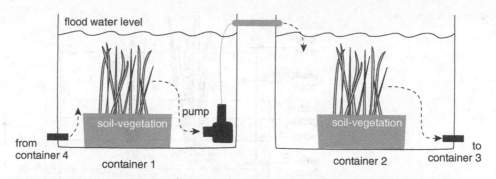

Figure 5-5 Mesocosm design used to investigate the effects of summer river flooding on plant diversity on soil-vegetation monoliths harvested from an impacted floodplain on the Rhine River in Germany and more pristine monolith harvested from the Narew River in Poland. (*After Antheunisse and Verhoeven, 2008*)

Florida Everglades

Mesocosms have been used frequently to determine management strategies and to investigate freshwater marsh ecosystem function in the Everglades in southern Florida. Some mesocosms there are associated with attempts to create thousands of hectares of restored wetlands along the northern border of the Everglades as phosphorus sinks to slow down continuing invasion by *Typha domingensis* into the Everglades. *Typha* has already replaced over 10,000 ha of native sawgrass (*Cladium jamaicense*) communities because of high nutrient conditions caused by upstream agricultural runoff. The South Florida Water Management District (SFWMD), in collaboration with other agencies, has already constructed 18,000 ha of new wetlands, known as stormwater treatment areas (STAs), to intercept the high-phosphorus runoff from adjacent agricultural areas (Chimney and Goforth, 2006).

White et al. (2004, 2006) developed a set of 12 freshwater marsh mesocosms consisting of 12 fiberglass-lined plywood tanks measuring 5.9 m long by 1.0 m wide by 1.0 m deep to simulate parts of the original STA-1. Each mesocosm had 30 cm of peat taken from near the stormwater treatment area (STA) freshwater marshes. Half of the mesocosms were planted with *Typha* sp., while the other half had submersed aquatic vegetation. Four treatments were used in the study: continuously flooding with emergent vegetation (*Typha*); periodic drawdowns with emergent vegetation; continuous flooding with submersed aquatic vegetation; and periodic drawdowns with submersed vegetation. While phosphorus was reduced in all of the mesocosm treatments regardless of vegetation type or flooding regime, submersed aquatic vegetation appeared to work better in reducing dissolved phosphorus. These results led to a policy of emphasizing submersed aquatic vegetation wetlands in the design of these treatment wetlands.

Wozniak et al. (2008) developed and tested transportable and reusable 2-m^2 1.6-m diameter mesocosms in the Florida Everglades freshwater marshes for quantifying nitrogen cycling by enrichment with ^{15}N within the Everglades itself (*in situ*)

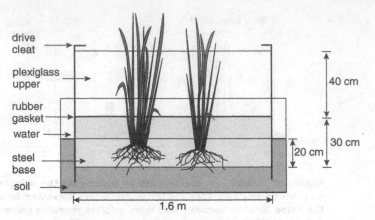

Figure 5-6   Design of reusable 1.6-m diameter *in situ* mesocosms used in Florida Everglades freshwater marshes for quantifying nitrogen cycling with nitrogen isotopes within the Everglades itself. Mesocosms include steel base connected to plexiglass upper structure. (*After Wozniak et al., 2008*)

(see Figure 5-6). Three of these mesocosms, after installation, were inoculated with ^{15}N isotope and three mesocosms were used as controls. Over a 21-day study period, Wozniak et al. (2008) found that all components of the ecosystem (macrophytes *Cladium jamaicense*, periphyton and consumers *Gambusia holbrooki*) except the soil itself had traces of the ^{15}N tracer. The quickest response was by the periphyton with later uptake by aboveground and then belowground *Cladium*. The researchers concluded that this mesocosm approach, because of the relatively large size (2 m^2) and short duration of the study, was not impacted by mesocosm artifacts such as pH and oxygen shifts. To run hydrologically isolated *in situ* studies for any longer period would probably introduce artifacts, due to the lack of replenishment of nutrients and needed water movement.

Peatland Mesocosms

Peatlands present a more difficult situation for mesocosm studies. They require excavation of large, intact monoliths of peat and surface vegetation for a reasonable representation of a bog or fen. Therefore, fewer peatland mesocosm studies are described in the literature. In the 1990s, Scott Bridgham and his colleagues developed a set of fifty-four 2.1-m^2 circular peatland monolith mesocosms (60 cm deep) at a research facility 70 km north of Duluth, Minnesota (see Figure 5-7a). The mesocosms have peat and surface vegetation extracted from both a fen and a bog in Minnesota while they were frozen in the winter (see Figure 5-7b). The overall experiment then had two types of peatlands, three infrared heating treatments, and three water table treatments (2 × 3 × 3) with three reps of each treatment combination. Most of the experiment was run in the mid- to late-1990s, with publications coming out a few years later (Bridgham et al., 1999; Weltzin et al., 2000; Updegraff et al., 2001;

Figure 5-7   a. Peatland study at a research facility 70 km north of Duluth Minnesota that included fifty-four 2.1-m^2 circular mesocosms (60 cm deep). b. The mesocosms had peat and surface vegetation extracted as monoliths from both a fen and a bog in Minnesota in the winter and inserted in the mesocosm tubs. The experiment had two types of peatlands, three infrared heating treatments, and three water table treatments with three reps of each treatment combination. (*Photos provided courtesy of S. Bridgham*)

Noormets et al., 2004). The authors found seasonal emissions of methane to be higher in the bog mesocosms than in the fen mesocosms and surmised that the lower rate of methanogenesis in the fens was due to higher methane oxidation in the aerated zones of these mesocosms (Updegraff et al., 2001). They also found that bogs and fens responded differently to warming and water table manipulations than might occur during climate shifts.

Whole-Ecosystem Experimental Wetlands

Whole-ecosystem studies are defined here as *in situ* ecological studies of a spatial and temporal scale that includes most if not all processes of that ecosystem. The importance of whole-ecosystem studies is more than just their large size. For example, a mesocosm experiment may investigate the effects of nutrients on plants but a whole ecosystem study is needed to investigate the role of nutrients on ecosystem functions, with plants being affected by the nutrients but also at the same time by herbivory, decomposition, sedimentation, and a host of other factors that are not independent of the nutrient inflow. A whole-ecosystem study purposefully does not simplify an ecosystem to derive cause and effect more easily; it attempts to include many more pathways and feedbacks in the system than do simpler systems (refer to Figure 5-1).

Theories now called self-organization, self-design, and ecosystem engineering, and practices such as ecological engineering, adaptive management, and ecosystem modeling emerged from whole-ecosystem studies. The impact of whole-ecosystem research on the teaching of ecology has been enormous. The principles of self-organization and self-design are key to whole-ecosystem studies and often do not occur at smaller space and time scales (Mitsch and Jørgensen, 2004). Whole-ecosystem research has been carried out by many investigators on a wide array of ecosystems, e.g., forests (e.g., Odum and Pigeon, 1970; Likens et al., 1977; Sullivan; 1993; Beier and Rasmussen, 1994) and lakes (e.g., Schindler, 1977, 1998; Schindler et al., 1997; Carpenter et al., 1996, 1998). There have also been a number of whole-ecosystem studies of wetlands (see Table 5-3). Some of those studies are discussed here.

Carolina Ponds

H. T. Odum has had a greater a role in whole-ecosystem studies than any other scientist. His experimental enclosures in Texas coastal bays, rain forest enclosures in Puerto Rico, created coastal ponds in North Carolina, and experimental cypress swamps in central Florida trained a generation of scholars on the importance of these studies. We presented that history in a tribute paper published after Odum's death in 2002 (Mitsch and Day, 2004) and some of that history is repeated here.

Odum (1985, 1989) reported on one of the first studies in the late 1960s that was done explicitly as an experiment in ecological engineering. Estuarine ponds in Morehead City, North Carolina, receiving secondarily treated municipal wastewaters mixed with salt water, were used to investigate ecological changes as the ponds

Table 5-3 Examples of Whole Ecosystem Experiments with Wetlands

Project	Location	Purpose	References
Experimental estuarine ponds	Morehead City, North Carolina	To investigate estuarine ponds receiving a mixture of wastewater and saltwater	Odum, 1985, 1989
Forested wetlands for recycling	Gainesville, Florida	To experimentally investigate forested cypress domes for wastewater recycling and conservation	Odum et al., 1977; Ewel and Odum, 1984; Dierberg and Brezonik, 1983a,b, 1985
Surface-water wetlands for wastewater treatment	Houghton Lake, Michigan	To use natural peatlands to treat wastewater from municipality to prevent lake pollution	Kadlec and Knight, 1996; Kadlec and Wallace, 2008
Marsh ecology research	Delta Marsh, Lake Manitoba, Canada	To investigate the effects of water level and water management on freshwater marsh ecology	Murkin et al., 1989; van der Valk et al., 1994
Restoration of riparian landscape	Des Plaines River Wetlands, Lake County, Illinois	To restore Midwestern U.S. river floodplain and determine design procedures for restored wetlands	Hey et al., 1989; Sanville and Mitsch, 1994; Mitsch et al., 1995;
Renovation of coal-mine drainage	Athens County, Ohio	To study iron retention from coal mine drainage using a *Typha* wetland	Mitsch and Wise, 1998
Nonpoint source pollution control	central Illinois created wetlands	To remove nutrients from Midwest agricultural runoff	Kovacic et al., 2000; Larson et al., 2000
Created flow-through riverine wetlands	Olentangy River Wetland Research Park, Ohio	To experimentally determine the long-term effects of planting on ecosystem function	Mitsch et al., 1998; 2005a,b; Anderson et al., 2005; Anderson and Mitsch, 2006
Restoration of peatland	Bois-des-Bel PeatlandQuébec, Canada	To monitor the restoration of a mined peatland	Rochefort et al., 2003; Isselin-Nondedeu et al., 2007; Waddington et al., 2003, 2008

developed. Three ponds received a sewage-saltwater mixture and three ponds were controls, receiving only a tap water–salt water mixture. All ponds were "seeded" with a high diversity of biotic communities from estuarine, freshwater, and sewage sources. Formally, the study focused on "whether the self-organization process [of species arrangements] occurs readily there with new conditions from wastewater influence and how much time is required" (Odum, 1989). The experiments demonstrated a rapid build-up of structure in the experimental ponds with heavy fringes of *Spartina* and blooms of the alga *Monodus* sp. in the fall and winter. The study concluded that while the estuarine wastewater ponds developed conditions that are often viewed as undesirable (e.g., algal blooms), they had organized ecological structure and could be valuable for the design of aquaculture systems or the use of ponds as natural tertiary treatment systems. Odum (1989) concluded that the design parameters such as inflows, loading rates, and other controls, along with the very important multi-seeding of as many species as possible, set the boundary conditions for a relatively rapid process of ecosystem development. This was essentially an experiment in self-organization and self-design.

Florida Cypress Domes

It is now common to see wetlands used for the treatment of wastewater (Kadlec and Knight, 1996; Mitsch and Jørgensen, 2004; Mitsch and Gosselink, 2007; Kadlec and Wallace, 2008). One of the first experiments to investigate the idea of recycling domestic wastewater into wetlands for improving water quality was carried out in the early 1970s in north-central Florida by H. T. Odum and his colleagues and students (Odum et al., 1977; Ewel and Odum, 1984). In a whole-ecosystem experiment, cypress domes (0.5–1.5 ha in size) dominated by pond cypress (*Taxodium distichum* var. *nutans*), pine, and hardwoods received high nutrient wastewater for several years from a trailer park near Gainesville, Florida (see Figure 5-8).

Two cypress domes received treated wastewater from the trailer park at a rate of about 2.5 cm week^{-1}, while a third dome received an equivalent of low-nutrient groundwater inflow and a fourth served as a natural control. There was more than 90% reduction in concentrations of nutrients, organic matter, and minerals by the wetlands receiving wastewater as the water passed to groundwater (Dierberg and Brezonik, 1983a,b, 1985). Nitrogen and phosphorus concentrations in the foliage and branches of the trees increased as wastewater was added and decreased again after the treatment stopped. This study also demonstrated that forested wetlands could be used in some cases to remove nutrients from wastewater with a minimum application of expensive and fossil energy-consuming technology (see Table 5-4). Using wetland ecosystems to solve human problems minimizes our fossil fuel and carbon footprint while protecting our valuable wetlands.

Delta Marsh, Manitoba, Canada

Ten experimental freshwater marsh cells referred to as the Marsh Ecology Research Program (MERP) were constructed within the Delta Waterfowl and Wetlands

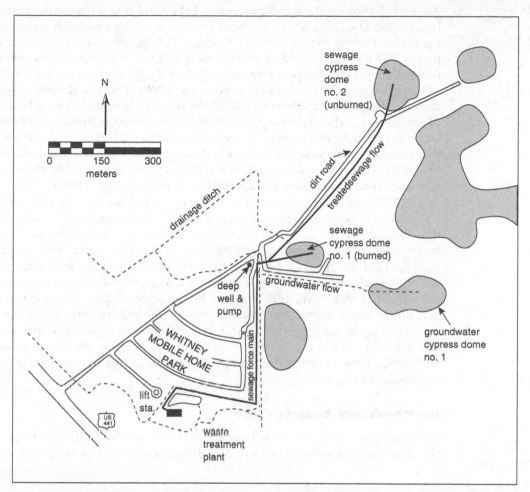

Figure 5-8 Whole-ecosystem wetland study site of cypress (*Taxodium*) domes in north-central Florida in the 1970s. Treated sewage wastewater from the mobile-home park was added to two cypress domes and an equivalent amount of low-nutrient groundwater was applied to a third cypress dome. (*After Odum et al., 1977*)

Table 5-4 Energy Costs of Tertiary Treatment in Florida Cypress (*Taxodium*) Swamps versus Conventional Technology. Energy Contribution from Nature, in Fossil Fuel Equivalents, Is Also Indicated for the Cypress Dome Treatment System

	Cypress dome treatment	Conventional tertiary treatment
Energy Cost, kJ m^{-3}	3,625	28,000
Natural energy subsidy, kJ m^{-3}	3,650	~ 0

Source: Adapted from Mitsch and Day, 2004.

Research Station on the southern shore of Lake Manitoba, Manitoba, Canada, in the late 1970s. The Delta Marsh has been a site of significant research on waterfowl in the Prairie Pothole region of North America. The specific rectangular cells that were part of MERP had an area of 6–8 ha each and had a gradient from permanently flooded to upland with open water (see Figure 5-9). Most of the studies conducted in these experimental marsh cells occurred in the 1980s. The four dominant vegetation communities in the experimental cells at the time of these studies were hybrid cattail (*Typha glauca*), whitetop rivergrass (*Scolochloa festucacea*), common reed (*Phragmites australis*), and hardstem bulrush (*Scirpus lacustris*). Murkin et al. (1989) found that there were significant differences in decay rate for some dominant macrophytes. *Scolochloa festucacea* and *Scirpus lacustris* decomposed more rapidly than did *Typha* spp. and *Phragmites australis* during their three-year study (1979–1982). Neckles et al. (1990) presented results from the same period on the influence of flooding on macroinvertebrate communities and found, not surprisingly, that permanent flooding reduced the density of invertebrates over natural hydroperiods. A replicated study using the 10 ponds investigated the effects that water level and flooding duration had on the wetland vegetation communities (van der Valk, 1994; van der Valk et al., 1994). The three water treatments in the experimental cells were: normal mean water level as determined by the adjacent lake, 30 cm above normal, and 60 cm above normal. As expected, plant richness, diversity, and shoot density decreased significantly in the flooded cells. Delta Marsh was also the site where van der Valk (1981) had previously gathered some of his data that supported his Gleasonian sieve model of wetland plant succession (see summary in Mitsch and Gosselink, 2007).

Des Plaines River Wetlands, Illinois

Restoration of entire rivers has been shown to be an elusive goal in many parts of the Midwestern United States because of significant loads of sediments and other

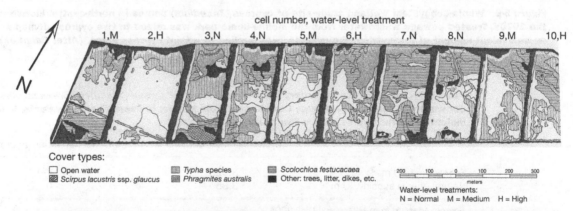

Cover types:

☐ Open water   ▦ *Typha* species   ▤ *Scolochloa festucacaea*
▨ *Scirpus lacustris* ssp. *glaucus*   ▩ *Phragmites australis*   ■ Other: trees, litter, dikes, etc.

Water-level treatments:
N = Normal   M = Medium   H = High

Figure 5-9   Major vegetation zones in the MERP complex at Delta Marsh, Manitoba, Canada. Cell numbers and water-level treatments are indicated above each cell. (*After van der Valk et al., 1994*)

non-point pollutants. In an ecological sense, we have paid too much attention to the stream itself and not enough to the interactions of the river with its floodplain. The Des Plaines River Wetland Demonstration Project, a brainchild of Donald Hey, involved restoration of a length of a river floodplain and establishment of experimental wetland basins on the floodplain where the dynamics of sediment and nutrient control could be determined in experimental fashion by using large pumps to introduce river water to the wetland basins. The project was located north of Chicago in Lake County, Illinois, and was initially assisted by a committee of researchers led by Bob Kadlec, University of Michigan; Bill Mitsch, The Ohio State University; and Arnold van der Valk, Iowa State University. Project goals were "to demonstrate how wetlands can benefit society both environmentally and economically, and to establish design procedures, construction techniques, and management programs for restored wetlands" (Hey et al., 1989).

The Des Plaines River research was carried out at two scales. On the entire 182-ha site, woody and scrub vegetation was replanted with native prairie species and oak savannas. Abandoned quarry lakes were connected to the river to give additional sediment trap efficiency and provide backwater habitats for fish and shorebirds. On the whole-ecosystem scale, four wetland basins (1.6–4.7 ha) were constructed and instrumented at the northern half of the site for precise hydrologic control to investigate the importance of hydrologic flow in otherwise similar wetlands (see Figure 5-10). Within this overall experimental design, several research questions were asked:

- Will water quality improvement and sediment retention be lower in wetlands with high-flow rates?
- Will the differences in the major forcing function (hydrologic flow-through) lead to different ecosystem development?
- Will experimental wetlands with higher hydrologic flows be more productive?
- Will the increased flowthrough bring substantially more sediments than can be assimilated, turning the subsidy to a stress?

Water was pumped through the wetlands at 7–16 cm per week (low-flow conditions) and 34–97 cm per week (high-flow conditions) with the basins maintained at similar depths (approximately 0.7 m average depth). Phosphorus retention was higher in high-flow wetlands (26–55 mg-P m^{-2} wk^{-1}) than in the low-flow wetlands (8–33 mg-P m^{-2} wk^{-1}) but percent retention of phosphorus was higher in low-flow wetlands (Mitsch et al., 1995). The high-flow wetlands maximized phosphorus mass retention, whereas the low-flow wetlands optimized phosphorus retention efficiency. The effects of flow conditions were investigated for hydrology (Hey et al., 1994a), water quality (Hey et al., 1994b), phosphorus dynamics (Mitsch et al., 1995), sedimentation (Brueske and Barrett, 1994; Fennessy et al., 1994a), aquatic metabolism (Cronk and Mitsch, 1994a), macrophyte productivity (Fennessy et al., 1994b), periphyton productivity (Cronk and Mitsch, 1994b), nitrogen dynamics (Phipps and Crumpton, 1994), atrazine fate (Alvord and Kadlec, 1996), and soil development

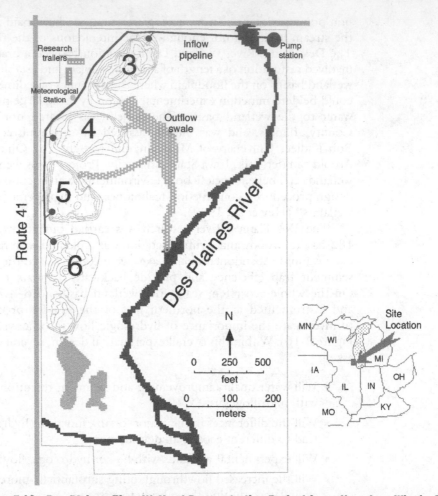

Figure 5-10 Des Plaines River Wetland Demonstration Project in northeastern Illinois. The site was part of a multi-year experiment with created freshwater marshes fed by riverwater from the Des Plaines River in the 1980s and 1990s. Research is described in several papers included in a special issue edited by Sanville and Mitsch (1994).

(Fennessy and Mitsch, 2001). The study found, for example, that after two years of experimentation, water column productivity was higher in high-flow wetlands than in low-flow wetlands but macrophytes productivity did not respond to the difference in hydrology; that latter effect might take decades to manifest itself.

Olentangy River Wetland Research Park, Ohio

The 20-ha Wilma H. Schiermeier Olentangy River Wetland Research Park (ORWRP), located on The Ohio State University's campus in Columbus, Ohio, USA, was started

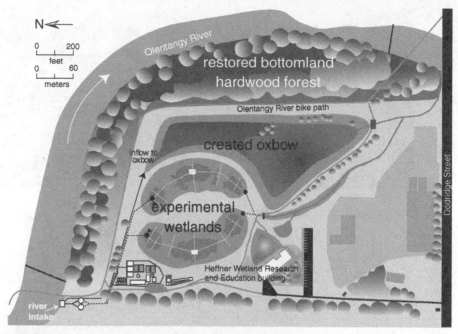

Figure 5-11   Wilma H. Schiermeier Olentangy River Wetland Research Park at The Ohio State University. This 21-ha wetland research facility includes two 1-ha experimental wetlands, a 3-ha created oxbow, and a 5-ha restored bottomland hardwood forest.

in 1993 as the first full-scale wetland research park on a university campus (see Figure 5-11). Fifteen years later, in June 2008, the site was named the 24th Ramsar wetland of international Importance in the USA, in recognition of the biological diversity of its created and restored wetlands as well as its record in university research and teaching, both to university students and the general public. (See Chapter 14, Mitsch and Gosselink, 2007 for a description of Ramsar wetlands of international importance). The goal of the ORWRP has been to provide teaching, research, and service on wetland science and wetland ecotechnology.

There are three main wetland creation and restoration projects on the site (see Figures 5-11 and 5-12), as follows:

- Two 1-ha flow-through experimental created wetlands (called experimental wetlands).
- A 3-ha created riparian oxbow wetland.
- A 5-ha restored bottomland hardwood forest along a 0.5-km stretch of the Olentangy River.

The site also has a mesocosm compound in its northwest corner with 80 installed experimental $1m^2$ mesocosms (described above) and a fifth set of larger flowthrough mesocosms. This "wetland campus" also has a visitor's pavilion where the public can

Figure 5-12   Olentangy River Wetland Research Park photos of a. 1-ha experimental wetland with board-walks and data collection system, b. 3-ha created oxbow, and c. 5-ha restored bottomland hardwood forest during winter flooding.

view the wetlands from above, a 3 kilometer bike path, a bikepath shelter with a solar energy collector on the roof, and a comprehensive wetland research and education building.

After the two 1-ha experimental wetlands and a river water delivery system were constructed in 1993–1994 at the ORWRP, over 2400 plant propagules (mostly root stock and rhizomes) representing 13 species typical of Midwestern U.S. marshes were planted in one wetland (Wetland 1, W1) in May 1994. Wetland 2 (W2) remained unplanted. Both wetlands have received the same amount and quality of pumped river water and remained essentially identical hydroperiods since pumping began in March 2004. River water is pumped into the wetlands continuously, day and night, except for occasional short-term unscheduled electrical failures. After start-up trials in 1994, a pumping protocol was developed that involves changing the pumping rate two or three times per week based on a formula that relates pumping rate to river discharge.

In 2003, extramural funding from Ohio and Federal agencies was obtained for a pulsing study whereby artificial "floods" were introduced to the wetland basins;

each wetland was administered with the same hydrologic conditions, so the "planting experiment" was not violated.

Pumped inflow to each wetland has averaged 20–30 m/year with water depths in the major portions of the wetland generally 20–40 cm in the shallow areas where most of the emergent macrophytes grow and 50–80 cm in the deeper areas that were created in the wetland to allow overwintering of fish (designed in the wetland for mosquito control) and long-term sediment retention. Early results illustrated an early system divergence and convergence of ecosystem function between the planted and unplanted wetlands (Mitsch et al., 1998). There was a clear pattern of ecosystem divergence six years after planting (Mitsch et al., 2005a) but the systems began to converge by the tenth year (Mitsch et al., 20005b). Overall, the experimental wetland that was planted continued to maintain a higher community diversity 14 years after planting, while the "unplanted" wetland that colonized naturally has usually been more productive and thus is accumulating more carbon (see Figure 5-13). Human planting appears to have enhanced macrophytes diversity; nature's planting enhanced ecosystem power.

Other studies published on the ecosystem-scale experimental wetlands at the ORWRP include those on aquatic system modeling (Metzker and Mitsch, 1997), algal dynamics and aquatic productivity (Wu and Mitsch, 1998; Tuttle et al., 2008), hydrology (Koreny et al., 1999; Zhang and Mitsch, 2005), water quality (Mitsch et al., 1998, 2005a,b; Kang et al., 1998; Nairn and Mitsch, 2000; Spieles and Mitsch, 2000a), benthic invertebrates (Spieles and Mitsch, 2000b, 2003), sedimentation (Harter and Mitsch, 2003; Nahlik and Mitsch, 2008), *Typha* hybridization (Selbo and Snow, 2004), methane generation (Altor and Mitsch, 2006, 2008b), denitrification and N_2O production (Hernandez and Mitsch, 2006, 2007a,b) and soil development (Anderson et al., 2005; Anderson and Mitsch, 2006).

The bottomland hardwood forest at the ORWRP, first described by Dudek et al. (1998) before its restoration, has been the result of several recent publications that have investigated sedimentation (Zhang and Mitsch, 2007) tree growth (Anderson and Mitsch, 2008a,b) and understory plant diversity (Swab et al., 2008) after restoration was undertaken by reintroducing river flooding in 2000–01 (see description of the floodplain restoration in Mitsch and Jørgensen, 2004).

Early results from the created oxbow wetland at the ORWRP were described by Mitsch and Jørgensen (2004) while the effectiveness of the oxbow in improving water quality with and without river pulsing was described by Fink and Mitsch (2007). This wetland responds automatically to the flood pulses of the Olentangy River and has a well-documented record of providing ecosystem services of flood mitigation, water quality improvement, and habitat development.

Bois-des-Bel Peatland, Québec, Canada

Despite the vast expanses of peatlands in the world, whole-ecosystem experiments on this type of wetland are rare. Bois-des-Bel peatland, located about 200 km northeast of Québec City, on the southern shore of the St. Lawrence River in Québec, Canada

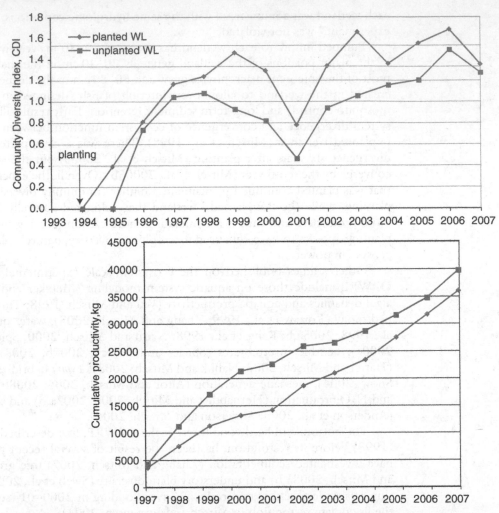

Figure 5-13 Olentangy River Wetland Research Park experimental wetland data since they were created in 1994: a. marsh vegetation community diversity, and b. cumulative marsh biomass production (since 1997) of planted and unplanted (naturally colonizing) basins. Wetland 1 was planted in May 1994, whereas Wetland 2 was not planted by humans.

(see Figure 5-14a), is a whole-ecosystem research site where scientists are evaluating the pace of peatland restoration after peat mining (Rochefort et al., 2003). The entire peatland is about 210 ha; the research area is about 11.5 ha of peatland that was drained in 1972 and mined by a vacuum extraction technique from 1973 to 1980. When mining stopped, a 2-m peat deposit remained (see Figure 5-14b). Restoration began in 1999 on 8.4 ha of the site, with the remaining as an unrestored control. The restored area (see Figure 5-14c) was divided into four zones, each of which has two

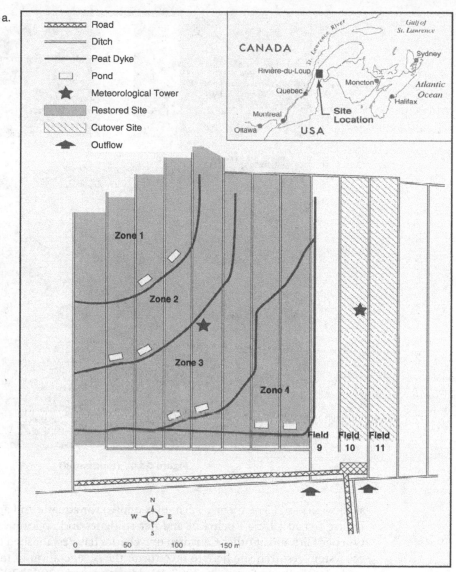

Figure 5-14 Bois-des-Bel experimental peatland on the southern shore of the St. Lawrence River in Québec, Canada: a. map showing restored (west) and cutover (east) sites including four research zones in restored area; b. non-restored section of the Bois-des-Bel site. Although this area was abandoned more than 25 years ago, plant recolonization is poor; c. visit of the Bois-des-Bel restored site seven years following restoration by researchers and members of the Canadian peat industry; d. vegetation monitoring in created pool at Bois-des-Bel restored site, eight years following restoration. Shallow pools were created in the restored area for aquatic and amphibian habitat. (*a. from Waddington et al., 2008, reprinted with permission; b., c., d, photos courtesy of PERG, Université Laval, Quebec City Valet, reprinted by permission.*)

Figure 5-14 (*continued*)

shallow pools (13 m x 5 m x 1.5 m max depth) for aquatic and amphibian habitat (see Figure 5-14d). Line Rochefort and her students and colleagues at Université Laval, Quebec City and at other Canadian universities have established the site as a long-term ecosystem research site here to investigate the revegetation of mined peatlands (Price et al., 1998; Rochefort et al., 2003; Waddington et al., 2003, 2008; Isselin-Nondedeu et al., 2007). The restoration involved terracing to produce better water distribution, reintroduction of *Sphagnum* diaspores harvested from a nearby natural wetland, and reflooding by blocking drainage ditches. The moss carpet increased by about 12 cm by 2007 and was three times the thickness that it was in 2003. *Sphagnum* cover by 2005 was 60% of the area in the restored sites compared to only 0.25% in the non-restored sites (Isselin-Nondedeu et al., 2007; see Figure 5-14b,c). The restored sites also exported less than half the dissolved organic carbon than did the cutover peatland sites (Waddington et al., 2008).

Mathematical Models

Mathematics is the language of science. When it is not possible to carry out all possible permutations of wetland behavior in a mesocosm or full ecosystem experiment, mathematical simulation models still allow an ecosystems perspective. While it is not our intention to present a modeling book here (there are several good ones, notably Odum and Odum, 2000; Jørgensen and Bendoricchio, 2001), we think it is appropriate to present some examples of simulation models of wetlands. There have been a great number of mathematical models of wetlands developed, particularly over the past 35 years. They range from generic models of salt marshes to a sophisticated spatial model of the vast Florida Everglades.

The translation of wetland ecosystems into a set of equations such as differential equations is foreign to a great many new wetland ecologists and that is unfortunate. They limit their ability to "see" ecosystems and instead find greater comfort in the mathematics of statistics, i.e. reductionist science. Models that describe and predict wetlands on a dynamic and spatial scale should be at the core of wetland ecosystem science—the emphasis of this book. Here we describe several wetland models in the familiar English language, and occasionally in the more descriptive systems diagrams (as introduced in Chapter 1). If readers wish to explore the actual mathematics of these models, they are invited to find the citations we provide for these models. As with the earlier part of this book, we divide the discussion on modeling into the few main types of wetlands covered in this book.

Tidal Salt Marsh Models

The first wetlands to receive the attention of mathematical modelers were salt marshes. A general conceptual model of a tidal salt marsh was presented in Figure 1-2. The first "model" (although not a computer simulation model) of a salt marsh was an energy budget of a Georgia salt marsh created by John Teal (1962) (see Figure 5-15a). It remains a classic attempt to quantify energy fluxes in the salt marsh, although many of the values might be modified today. Gross primary productivity was 6.1% of incident sunlight energy, verifying the observation that salt marshes are one of the most productive ecosystems in the world. Herbivorous insects, mostly plant hoppers *(Prokelisia)* and grasshoppers *(Orchelimum)*, consumed only 4.6% of the *Spartina* net productivity, with the rest of the *Spartina* productivity and that of the mud benthic algae passing through the detrital-algal food chain or exporting to the adjacent estuary. An estimated 45% of the net production was exported from the marsh into the estuary in this study.

Reimold (1974) expanded on the phosphorus model of a *Spartina* salt marsh developed by Pomeroy et al. (1972) in one of the first simulation models of salt marshes. The model has five phosphorus compartments: water, sediments, *Spartina*, detritus, and detrital feeders, and it was used to simulate the effects of perturbations such as *Spartina* harvesting on the ecosystem and to help design subsequent field experiments. The simulations demonstrated that *Spartina* regrowth and the resulting phosphorus in the water depended on the time of year that the harvesting of the marsh grass took place. Morris and Bowden (1986) developed a more detailed model of the sedimentation of exogenous and endogenous organic and inorganic matter, decomposition, aboveground and belowground biomass and production, and nitrogen and phosphorus mineralization in an Atlantic Coast salt marsh. They found that the model calculations of nitrogen and phosphorus export were sensitive to small changes in belowground production and the fraction of refractory organic matter, both parameters for which there is little good verification. The model demonstrates that mature marshes with deep sediments recycle proportionally more of the nutrients required for plant growth than do young marshes.

Dick Wiegert and his associates (Wiegert et al., 1975; Wiegert et al., 1981; Wiegert and Wetzel, 1979; Chalmers et al., 1985; Wiegert, 1986) put considerable effort into the development of a simulation model for the salt marshes of Sapelo Island, Georgia. Their model (see Figure 5-15b) had 14 major compartments and traces the major pathways of carbon in the ecosystem through *Spartina*, algae, grazers, decomposers, and several compartments of abiotic carbon storage. The model was originally constructed to answer three questions:

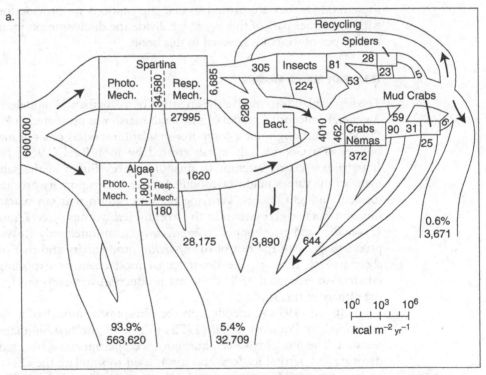

Figure 5-15   Salt marsh models of Sapelo Island Georgia salt marsh: a. John Teal's energy flow diagram; b. Dick Wiegert's simulation model. (*a. from Teal, 1962; b. from Wiegert et al., 1981*)

b.

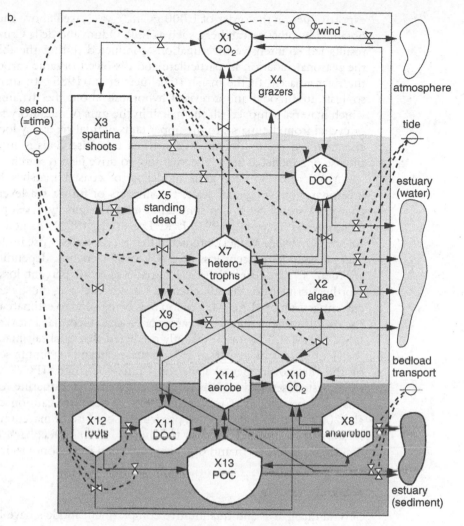

Figure 5-15 (*continued*)

- Is the Georgia salt marsh a potential source of carbon for the estuary, or is it a sink for carbon from offshore?

- What organism groups are most responsible for the processing of carbon in the salt marsh?

- What parameters are important (but poorly known) for the proper modeling of the salt marsh? (Wiegert et al., 1981).

One of the most important revelations from this model-building process was the demonstration of the importance of the tidal export coefficient. It consistently gave export values on the order of $1,000$ gC m^{-2} yr^{-1}, well above the field measurements obtained by other researchers, such as E. P. Odum and de la Cruz (1967). The model results led to intensive investigations conducted during the following five years of the seasonal variation in particulate and dissolved organic carbon concentrations in the river adjacent to the marsh (Chalmers et al., 1985). In turn, this led to a more sophisticated model and a revised hypothesis about the dynamics of carbon flux in which material is moved off the marsh in the guts of migratory feeding fish and birds or cycled from the marsh to the upper ends of tidal creeks by local rainfall at low tide and then redeposited on the marsh when tides rise. This is an excellent example of the use of a model as an intellectual tool to drive field research.

More recently, ecological modeling of coastal marshes has been applied to investigate, on a regional scale, the impacts of future sea level rise due to climate change. A spatial cell-based simulation model originally developed by Dick Park and colleagues (Park et al., 1991; J. K. Lee et al., 1991), named SLAMM (Sea Level Affecting Marshes Model), predicted that a 1 m sea level rise in the next century could lead to 26 to 82 percent loss of U.S. coastal wetlands, depending on the protection afforded. J. K. Lee et al. (1991) predicted a 40 percent loss of wetlands of the Northeastern Florida coastline with a 1 m sea level rise, mostly as low salt marsh. The current version of SLAMM, Version 5, is being used to estimate the effects of sea level rise on salt marshes and coastal areas of Florida, Georgia, and Washington state in the USA. SLAMM, Version 5, recently predicted that tidal salt marshes will decrease on the Georgia coastline by 20 to 45% for the mean and maximum sea level rises predicted by the Intergovernmental Panel on Climate Change (IPCC) with subsequent loss of ecosystem services such as organic export to the estuarine food chains and water quality improvement (Craft et al., 2009). There was caution expressed by Craft et al. (2009) in the model results, particularly because of inaccurate land elevation data in some locations and because the model lacks any feedback mechanisms, such as increased plant growth and organic production in response to increased sea levels.

Mangrove Models

Several qualitative and quantitative compartment models have been developed from research on the functional characteristics of mangroves in south Florida. In our simple conceptual model of energy and material flows (refer to Figure 1-4), the mangrove ecosystem is shown to be affected principally by tidal exchange and upland inflow but also internally by the role of crabs capturing fresh litter and consuming and burying it, thus reducing the amount of outwelling. Occasional hurricane pulses turn live mangroves into standing dead wood, but the effects are temporary rather than long lasting.

Chen and Twilley (1998) described a simulation model that predicts biomass and composition of mangrove forests through time (see Figure 5-16). This model tracks the growth of each tree in a forest gap of defined size, based on species-specific life-history traits and limitations of resource availability on the individual. This approach has been used extensively to model temperate and boreal forests and to simulate terrestrial forest dynamics. Only three mangrove species occur in south

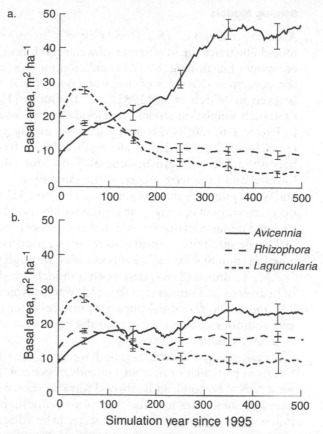

Figure 5-16   Projected basal area of three mangrove species at the lower Shark River estuary, Florida, simulated with a model under different recruitment rates: a. equal recruitment rate for the three mangrove species; b. low black mangrove (*Avicennia germinans*) recruitment rate. (*After Chen and Twilley, 1998*)

Florida, although the three species can form diverse ecological types—riverine, fringe, basin, overwash, hammock, and dwarf mangroves (see Chapter 2). Simulations indicated that the initial conditions of the forest following disturbance, recruitment, and nutrient availability acted in combination to control mangrove regeneration. The model also predicted that *Laguncularia racemosa* (white mangrove) would dominate the early stages of forest development because it can resprout after trunk destruction, and grows faster than the other two species in fertile and low salinity environments (see Figure 5-16a). However, in 500-yr simulations the early dominance of *L. racemosa* was replaced by *Avicennia germinans* (black mangrove), even when the recruitment rate of the latter was reduced to half that of the other species (see Figure 5-16b). Models such as this are useful in forest studies, including mangroves, because of the impossibility of obtaining long-term data in most sites.

Swamp Models

A conceptual model of a Florida cypress swamp was shown in Figure 1-6. This model illustrates the importance of water level, fire, nutrients, and tree harvesting on ecosystem functioning. Several simulation models have also been developed to study the ecosystem dynamics of deepwater swamps. A review of some of these models is given in Mitsch et al. (1982), H. T. Odum (1983, 1984), and Mitsch (1988). One such simulation model of a Florida cypress swamp, similar to the model shown in Figure 1-6, was used to investigate the management of cypress domes in Florida (Mitsch et al., 1982). The model predicted long-term (100 years) effects of drainage, harvesting, fire, and nutrient disposal. Simulations showed that when water levels were lowered and trees were logged at the same time, the cypress swamp did not recover and was replaced by shrub vegetation. The model also suggested that if tree harvesting occurred without drainage, the cypress would recover because of the absence of fire. In a related minimodel that investigated understory productivity in cypress domes, Mitsch (1984) demonstrated annual patterns of aquatic productivity that peaked in the spring when maximum solar radiation is available through the deciduous cypress canopy.

M. T. Brown (1988) developed a model that demonstrated the annual patterns of hydrology and nutrients in forested swamps typical of central and northern Florida. The model illustrated the impact of surface water diversion, lowering groundwater, and additions of advanced wastewater treatment effluents on the hydroperiod and nutrient dynamics of the swamps.

Few energy-nutrient models have been developed for bottomland hardwood forests in particular or riparian wetlands in general. A general energy/nutrient model for a typical bottomland hardwood forest is shown in Figure 1-7. The flux of a river past and sometimes into the floodplain (longitudinal exchange) and the important inflow of groundwater and surface water from adjacent uplands (transverse exchange) typify riparian ecosystems. The primarily detrital food chain includes a major export of detrital material into the river, particularly during flooding. The water, nutrients, and sediments that are transported into the bottomland forest during flooding represent the energy subsidy that makes these systems among the most productive forested systems in a given region. The scarcity of specific energy-nutrient simulation models can be attributed as much as any cause to the difficulty in quantifying the relationships between stream flooding and ecosystem productivity.

Transition models based on individual tree growth and their responses to the environment include a suite of models based on SWAMP, a tree-growth simulation model of southern wetland forests. This model was developed by Phipps (1979) and applied to a bottomland forest in Arkansas and later to a forested wetland in Virginia (Phipps and Applegate, 1983). As with the mangrove model described above, this model simulates the growth of individual trees in the forest, summing the growth of all trees in a plot to determine plot dynamics. The model includes subroutines, including GROW, which "grows" trees on the plot according to a parabolic growth form; KILL, which determines the survival probabilities of trees and occasionally "kills" trees; and CUT, which enables the modeler to remove trees, as in lumbering or insect damage. Another important subroutine, WATER, describes the influence of water-level fluctuations on tree growth. This subroutine assumes that tree growth will be suboptimal during the peak May-June growth period if water is either too high or too low. The effect of water levels on tree productivity is hypothesized to be a parabolic function as follows:

$$H = 10.05511(T - W)^2 \tag{5.1}$$

where

H = growth factor related to the water table,

T = water table depth of sample plot, and

W = optimum water table depth.

The model begins with all tree species greater than 3 cm in diameter on a 20 m × 20 m plot and "grows" them on a year-by-year basis, generating results that depend on hydrologic conditions such as flooding frequency and depth to the water table, and other factors such as shading and simulated lumbering. Typical results from runs of the model based on data from the White River in Arkansas indicate the importance of altered hydroperiod and lumbering for the structure of the bottomland forest. FORFLO, a model based on SWAMP, was applied to evaluate the impact of an altered hydrologic regime on succession in a bottomland forest in South Carolina (Pearlstein et al., 1985; Brody and Pendleton, 1987).

Mitsch (1988) developed a simple STELLA (see BOX below for description of STELLA) simulation model that illustrated the simultaneous effects of river flood pulsing and nutrient dynamics on forested wetland productivity (see Figure 5-17). Model simulations illustrated that productivity (net biomass accumulation) ranged from highest to lowest in the following order of hydrologic conditions:

pulsing > flowing > stagnant

The pulsing system had the highest productivity because of a combination of high nutrients and an optimum hydrology that allowed the water level to decrease between floods to levels closer to the optimum requirements of the vegetation. This model is a good illustration of the flood pulse concept (FPC) of floodplain wetland systems.

Modeling with STELLA

STELLA is a high-level simulation software with an icon-based graphical interface that simplifies modules and can be used easily for modeling and simulating wetland ecosystems without program coding. STELLA provides good opportunities to explore "what if" for modeling of wetland ecosystems over time. STELLA v. 9.1 has a powerful web link. Figure 5-18 provides the basic STELLA icons (codes) for describing modeling variables and processes for systems and compares them to the Odum energy language. For wetland ecosystems, for example, storages (e.g., water, dissolved oxygen, phosphorus, vegetation biomass) are the components to be modeled (dependent variables). Processes within the wetland ecosystem (e.g., water flow, productivity, nutrient cycling) and ecosystem force functions (e.g., solar energy, precipitation, tides) are also easily described in this symbolic language. If the diagram is constructed properly, the differential equation mathematics

(continues)

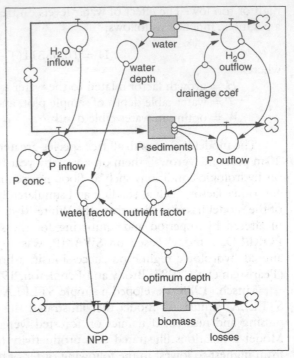

Figure 5-17 Simulation model in STELLA language of forested wetland. The model was simulated for low and high nutrient conditions and for stagnant, flowing, and pulsing hydrologic conditions. (*After Mitsch 1988*)

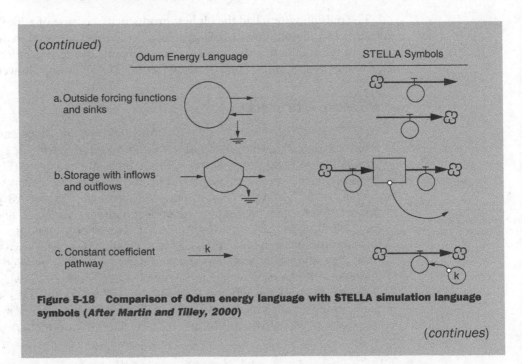

Figure 5-18 Comparison of Odum energy language with STELLA simulation language symbols (*After Martin and Tilley, 2000*)

(*continues*)

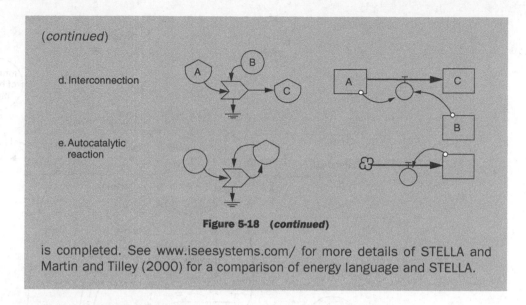

(continued)

d. Interconnection

e. Autocatalytic reaction

Figure 5-18   *(continued)*

is completed. See www.iseesystems.com/ for more details of STELLA and Martin and Tilley (2000) for a comparison of energy language and STELLA.

Freshwater Marsh Models

An ecosystem model of a freshwater marsh was illustrated in Figure 1-5. Nutrient cycling and food webs are complex because of the important roles of algae (periphyton and sometimes phytoplankton) as well as macrophytes as major plants and a wide variety of consumers, including benthos, amphibians, muskrats, fish, and waterfowl and other birds. This model shows the importance of the detrital pathways although some grazing of live plants occurs with zooplankton, fish, and muskrats.

Several models have investigated the dynamics of freshwater marshes, whether for predicting wildlife use or nutrient dynamics. Oliver and Legovic (1988) used a simulation model to evaluate the importance of nutrient enrichment in a freshwater marsh near a wading bird rookery in the Okefenokee Swamp in Georgia. The trophic level model illustrated the influence of the avian fauna on several aspects of the ecosystem, including detritus, macrophytes, and aquatic fauna. Benthic detritus increased 8.9 times, macrophytes 4.5 times, and fish 1.4 times the background levels as a result of the simulated influx of 8,000 wading birds into the marsh. Özesmi and Mitsch (1997) developed a habitat model that predicted the spatial distribution of breeding sites for the red-winged blackbirds (*Agelaius phoeniceus*) in marshes along Lake Erie in northern Ohio. The model concluded that nest site locations were most closely related to water depth and vegetation durability.

A series of simulation models have been developed at the Olentangy River Wetland Research Park since the early 1990s that emphasize freshwater marshes that are directly connected to creeks and rivers. Almost all models are lumped models, that is, they are not spatially explicit. They were all developed as first-order differential equation models and run on various versions of the software STELLA. These models were meant to emphasize ecological processes in wetlands, generally nutrient cycling, energy flow, or community dynamics.

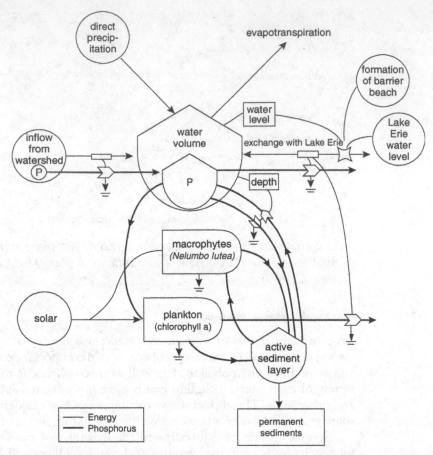

Figure 5-19 Simulation model for Old Woman Creek Wetland, a coastal freshwater wetland on Ohio's shoreline with Lake Erie. The model emphasized phosphorus dynamics and retention in the wetland and exchanges from the upstream watershed and with Lake Erie at the downstream end. (After Mitsch and Reeder, 1991)

An early generation freshwater marsh model that emphasized phosphorus cycling and ecosystem energetics (see Figure 5-19) was developed by Mitsch and Reeder (1991) for a coastal marsh along Lake Erie. It was used to estimate the phosphorus retention capability of this wetland under different hydrologic conditions. The model predicted, for example, that phosphorus retention was highest when watershed inflow and lake levels were both high. The model was revised as a second generation model to include more detail on the phosphorus component in the water column by Christensen et al. (1994) and applied to the experimental wetlands at the Des Plaines River Wetland Demonstration Project described above to simulate phosphorus retention under different hydrologic conditions.

A third generation of the Mitsch and Reeder model was improved significantly to compare wetland function and nutrient retention in low-flow and high-flow conditions already imposed on the Des Plaines River wetlands in the early 1990s

(Wang and Mitsch, 2000). The new model (see Figure 5-20a), which included four submodels—hydrology, primary productivity, sediments, and phosphorus—was robust, as it was calibrated and validated with three years of data from the four wetland basins at the Des Plaines site (refer to Figure 5-10) with the different hydrologic conditions. The validated model was used to estimate internal nutrient fluxes that otherwise could never be determined through field techniques such as

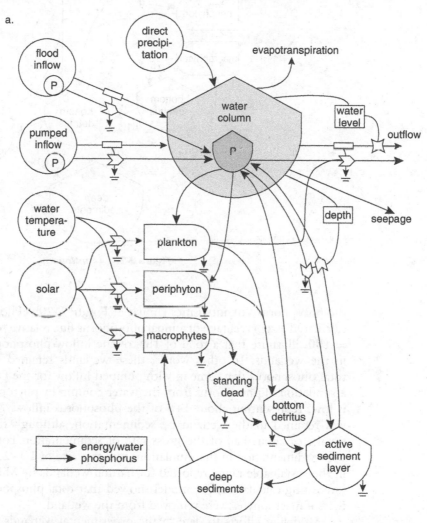

Figure 5-20   Simulation model used for created freshwater marshes at the Des Plaines River Wetland Demonstration Project, northeastern Illinois. a. energy flow diagram of the marsh model, and b. phosphorus flow intensities for the Des Plaines River wetlands as standardized to a phosphorus inflow of 100%. Phosphorus flow intensities are averaged from the model for all four wetlands simulated from October 1989 to September 1991. (*After Wang and Mitsch, 2000*)

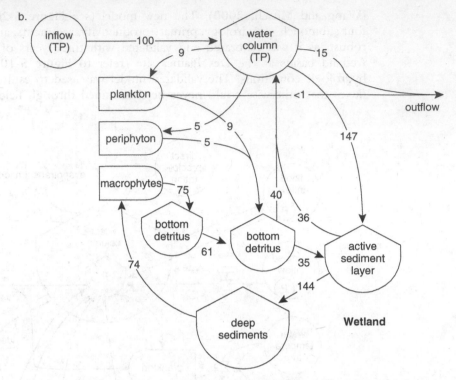

b.

Figure 5-20 (*continued*)

the phosphorus flux intensities shown in Figure 5-20b. Phosphorus flux intensities, calculated as a percentage of annual phosphorus flux relative to the phosphorus inflow = 100, illustrate that a total of 15% of the inflow phosphorus will not be retained in the wetlands. In other words, these wetlands retained on average 85% of the total phosphorus that came in with pumped inflow for the first two years. Biological assimilation of phosphorus from the water column by phytoplankton and periphyton is low, representing about 14% of the phosphorus inflow. Most of the phosphorus was retained in the wetland by sedimentation, although the macrophytes had an uptake of about half of the gross rate of sedimentation. Total phosphorus retained with sedimentation in the simulations was at a rate of 1.1–2.5 g-P m^{-2} yr^{-1}, similar in the sustainable range reported for natural wetlands by Mitsch et al. (2000). In an interesting calculation, the model showed that total phosphorus retention increased 5.1% if macrophytes were removed from the wetland.

Modeling efforts to date in the experimental wetlands at the Olentangy River Wetland Research Park have emphasized fish community dynamics (Metzker and Mitsch, 1997), wetland hydrology (Zhang and Mitsch, 2005), and aquatic productivity (Tuttle et al., 2008). The fish community model (see Figure 5-21) was run under a series of different environmental and biological conditions to predict the fish

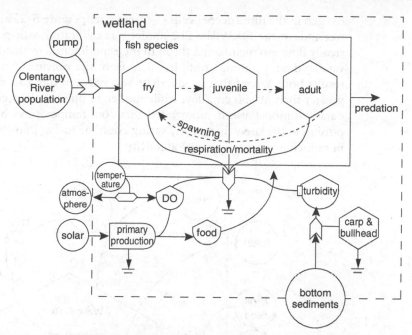

Figure 5-21   Conceptual diagram of aquatic community model simulated for experimental wetlands in Olentangy River Wetland Research Park soon after the construction of the wetlands. (*After Metzker and Mitsch, 1997*)

community structure in the experimental wetlands at the ORWRP. Overall, depending on the assumptions, simulations of the fish community over 10 years went in two different directions with the most probable a system dominated by carp (*Cyprinus carpio*) and supporting large biomass 500–930 kg/ha. The model showed that the fish community is not randomly selected for, but is a product of, self-design. Further, the model showed that it sometimes takes several years for fish populations to develop, even when all of the propagules are present.

Zhang and Mitsch (2005) presented the results of hydrologic simulation models developed for three of the wetland systems at the Olentangy River Wetland Research Park: the experimental wetlands that receive pumped river water, the 3-ha created oxbow that is fed by river flood pulses, and a small (0.13 ha) stormwater wetland that receives runoff from the research building roof during rain storms. Simulations of the experimental wetlands provided estimations of their average retention time of four to five days. The simulation of the created oxbow showed that it has more than enough flooding from the river in most years for it to have standing water sufficient for a formal jurisdictional wetland in the United States. The water budget for the stormwater wetland predicted hydroperiods for wet, average, and dry years. Model results compared well with later experience with these wetlands.

An aquatic productivity model of the experimental wetlands at the ORWRP (Figure 5-22a; Tuttle et al., 2008) predicted gross primary productivity and dissolved oxygen in the three deeper cells (>60 cm deep; Figure 5-22b) of the two experimental wetlands at the ORWRP. The model was calibrated with data from a year in which steady flow was maintained through the experimental wetlands (see Figure 5-22c) and validated with pulsing condition data from the previous year. Field data and model results both agreed that productivity was reduced due to a flushing effect with flows greater than 30–50 cm/day, while moderate pulses enhanced productivity. The field data and model also demonstrated that cool temperatures brought in by hydrologic pulsing in the late winter/early spring could be just as important as the flushing effect in reducing water column productivity.

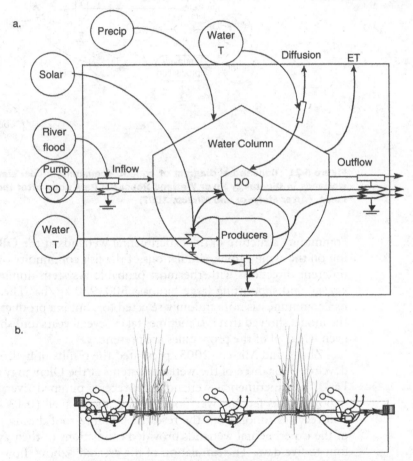

Figure 5-22   Water column metabolism model developed for deep pools in the Olentangy River Wetland Research Park experimental wetlands W1 and W2: a. conceptual model, b. spatial pattern of three wetland pools in sequence from inflow to middle to outflow, and c. model calibration results for March through June 2004 for gross primary productivity (GPP) in inflow, middle, and outflow of the wetlands compared to field measurements. (*After Tuttle et al., 2008*)

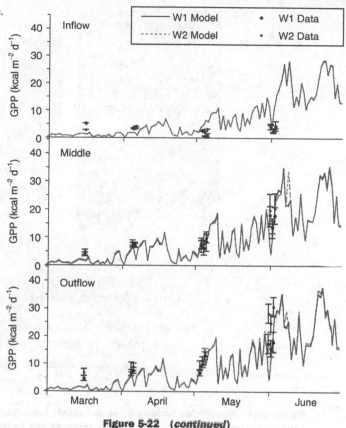

Figure 5-22 *(continued)*

Everglades Landscape Model

It is clear that much can be learned when the power of dynamic modeling is coupled with the mapping capabilities of GIS databases. Spatial models have been applied to understanding patterns of landscape fragmentation, invasion of undesirable vegetation, and ecosystem dynamics. Wu et al. (1996) developed a spatial model to understand the dynamics of fire in the Florida Everglades. They found that when *Typha* invaded the Everglades, the area burned and fire frequency decreased by 23 percent and 21 percent respectively. Deeper water in the Everglades has decreased the frequency of fire by 63 percent, making less frequent fires during drought seasons more severe. Cattail (*Typha*) expansion into the sawgrass (*Cladium jamaicense*) of the Florida Everglades was investigated with another spatial model that predicted that cattails would invade 50 percent of a major (43,000 ha) water conservation area in the Everglades in six to ten years if nutrient loading remained the same.

More recently, the Everglades Landscape Model (ELM) (Fitz et al., 2003; Figure 5-23) was an ambitious attempt to use a spatially explicit dynamic model of the Florida Everglades to be able to better understand this wetland's dynamics, to

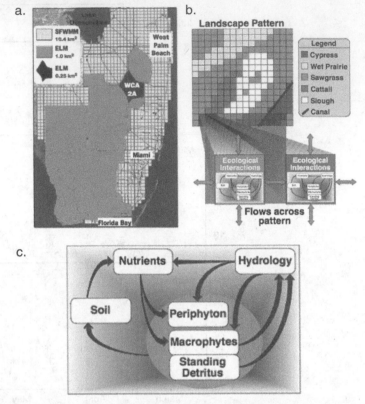

Figure 5-23 **Everglades Landscape Model (ELM) including a. model domains in southern Florida, b. spatial interactions that show connections between adjacent cells, and c. general conceptual model of major components of models in each cell. (*From Fitz et al., 2006*)**

be able to predict landscape response to hydrologic and nutrient changes, and to provide a tool for the ongoing Everglades restoration. The model was developed from the mid-1990s to 2007 at the South Florida Water Management District. The Everglades Landscape Model (ELM) is one of a family of regional ecosystem models in a simulation framework developed by Robert Costanza and his colleagues (see Costanza and Voinov, 2004). These landscape models have been implemented and parameterized for specific regions such as the Everglades, Patuxent, Gwynns Falls, Baltimore, Louisiana Delta, as well as the Florida Everglades. They are designed to link short- to long-term ecosystem, hydrologic and biogeochemical processes and factors into a systems-level approach to simulate material fluxes and storages over dynamic landscapes. Time scales can range from daily to decadal, and spatial scales typically range from 10^2 to 10^4 km^2. The main problem with a model of such spatial scale and complexity is that it can truly never be validated to be acceptable in its predictions. As with other modeling efforts, its main contribution may be to give an organized system vantage point for the Florida Everglades to better design field monitoring and wetland process research.

Peatland Models

There has been much research on peatland hydrology and how peatland hydrology and peatland carbon dynamics fit in climate change. Yet with the current interest in climate change and with wetlands in the position of linchpins in the carbon cycle of the planet (see Chapter 10, Mitsch and Gosselink, 2007), there has been relatively little work done on ecosystem models of carbon cycling in peatlands. Y. Zhang et al. (2002) present a relatively comprehensive model, Wetland-DNDC that was designed to predict both methane and carbon dioxide emissions from wetlands in the context of a model that includes hydrology, soil biogeochemistry, and wetland vegetation processes. The four main interacting components of the model are hydrology, soil thermodynamics, plant growth, and soil carbon dynamics (see Figure 5-24). The model was tested at three peatland sites in Saskatchewan and Minnesota where extensive data on methane fluxes in the 1990s were available. The model predicted the seasonal patterns of methane emissions reasonably well.

In Chapter 4, we introduced a detailed organic carbon flux diagram of a peatland in Russia (refer to Figure 4-17b) that was quantified years ago by Bazilevich and Tishkov (1982) and further refined by Alexandrov et al. (1994). Zavalishin (2008) explored the stability of these same transitional bogs in Tajozhny Log in the Novgorod Region of Russia. His model was simple; it had four main compartments—plants, animals, fungi, and bacteria, along with dead organic matter and litter. The simulations, over several hundreds of years, show complex oscillatory dynamics that makes describing the role of large expansive transitional bogs in climate change rather difficult. As summarized by the author:

> Elevation of the atmospheric CO_2 concentration yields an increase of carbon assimilation by bog's vegetation that results in a qualitative change of the carbon cycle dynamic behaviour—transfer firstly to periodic and then to chaotic oscillations of carbon content in compartments. This result confirms an evidence of the limited opportunity for bog ecosystems of a transitional (mesotrophic) type to accept carbon abundancies from the atmosphere without change of their qualitative behaviour. In contrast, decrease of the atmospheric carbon assimilation moves the ecosystem to the state of a raised bog, and if the output rate from the litter increases (due to either strengthening of peat formation or anthropogenic, or fire activities in litter destruction), then a meadow or a sphagnum pinery can arise. Intensification of dead biomass and litter decay transform the ecosystem in a state of a fen with neglectable amounts of carbon in animal and destructors pools. (Zavalishin, 2008)

Only through the use of an ecosystem model would such a complex description of the potential role of peatlands in our changing climate be possible.

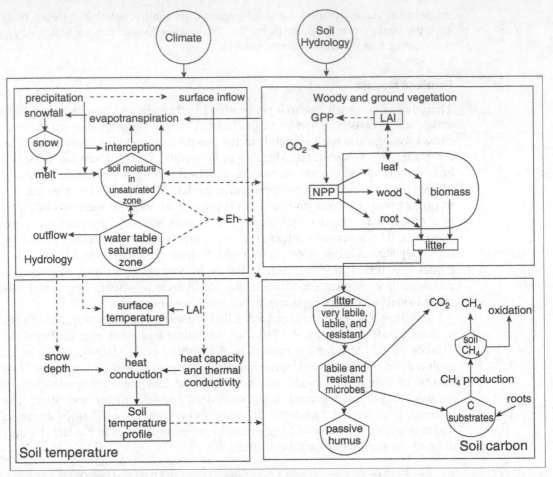

Figure 5-24 Wetland-DNDC model used to simulate carbon dynamics in northern peatlands. Four components were originally called hydrologic conditions, plant growth, soil thermal conditions, and soil carbon dynamics. Odum energy symbols were added to assist with interpretation of model. (*After Zhang et al., 2002*)

Conclusions

None of the systems approaches to studying wetland described here will allow us to determine all of the relationships governing wetland processes. But a holistic approach to studying systems in general and wetlands in particular gives an uncommon view that can be both enlightening and organizational. As stated in a previous book "reductionistic experiments, when run at small scales, can lead to erroneous management decisions if the results are extrapolated to full-scale ecosystems—the very level at which ecological engineering and restoration often work" (Mitsch and Jørgensen, 2004). But there is not a perfect systems approach to work within wetland science. Mesocosms give us replicated behavior of simplified wetland systems in relatively short experimental times. Whole ecosystem studies are more realistic, yet are enormously costly to maintain and are difficult to replicate. Although mathematical models of wetlands are simply an integrated set of algorithms, a more complex mathematics does not necessarily mean the model is closer to truth. All of these approaches encourage the wetland researcher to think in a holistic way, to sift through a multitude of possibilities to identify what is most important in the ecosystems being studied, and to serve as the integrator for those who are focused on the parts. Those are their true values. The whole remains greater than the sum of the parts.

Recommended Readings

Costanza, R. and A. Voinov, eds. 2004. Landscape Simulation Modeling: A Spatially Explicit, Dynamic Approach. Springer, Secaucus, New Jersey. 330 pp.

Jørgensen, S. E. and G. Bendoricchio. 2001. Fundamentals of Ecological Modelling, 3rd edition, Elsevier, Amsterdam. 530 pp.

Kangas, P. C. 2004. Ecological Engineering: Principles and Practice. Lewis Publishers, CRC Press, Boca Raton, Florida. 452 pp.

Mitsch, W. J. and S. E. Jørgensen. 2004. Ecological Engineering and Ecosystem Restoration. John Wiley & Sons, Inc., New York. 411 pp.

Odum, H. T. and E. C. Odum. 2000. Modeling for All Scales: An Introduction to System Simulation. Academic Press, San Diego. 458 pp.

Odum Energy Language Symbols

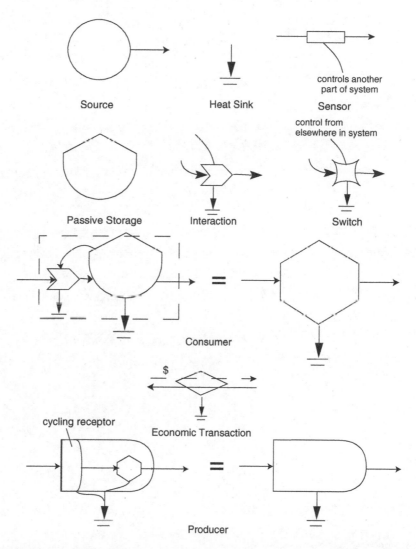

Source

Heat Sink

Sensor

controls another
part of system

control from
elsewhere in system

Passive Storage

Interaction

Switch

Consumer

$

Economic Transaction

cycling receptor

=

Producer

(From Mitsch and Jørgensen, 2004, reprinted with permission, John Wiley & Sons, Inc.)

References

Abernethy, R. K. 1986. Environmental Conditions and Waterfowl Use of a Backfilled Pipeline Canal. M.S. Thesis, Louisiana State University, Baton Rouge. 125 pp.

Abernethy, V., and R. E. Turner. 1987. U.S. forested wetlands: 1940–1980. BioScience 37: 721–727.

Adam, P. 1998. Australian saltmarshes: A review. In A. J. McComb and J. A. Davis, eds. *Wetlands for the Future*. Gleneagles Publishing, Adelaide, Australia, pp. 287–295.

Adey, W. H., Finn, M., Kangas, P., Lange, L., Luckett, C., Spoon, D. M. 1996. A Florida Everglades mesocosm—model veracity after four years of self-organization. Ecological Engineering 6: 171–224.

Ahn, C., W. J. Mitsch, and W. E. Wolfe. 2001. Effects of recycled FGD liner material on water quality and macrophytes of constructed wetlands: A mesocosm experiment. Water Research 35: 633–642.

Ahn, C., and W. J. Mitsch. 2001. Chemical analysis of soil and leachate from experimental wetland mesocosms lined with coal combustion products. Journal of Environmental Quality 30: 1457–1463.

Ahn, C., and W. J. Mitsch. 2002a. Evaluating the use of recycled coal combustion products in constructed wetlands: An ecologic-economic modeling approach. Ecological Modelling 150: 117–140.

Ahn, C., and W. J. Mitsch. 2002b. Scaling considerations of mesocosm wetlands in simulating large created freshwater marshes. Ecological Engineering 18: 327–342.

Alexandrov, G. A., N. I. Bazilevich, D. A. Logofet, A. A. Tishkov, and T. E. Shytikova. 1994. Conceptual and mathematical modeling of mater cycling in Tajozhny Log Bob ecosystem (Russia). In B.C. Patten, ed. *Wetlands and*

Shallow Continental Water Bodies, Vol. 2. SPB Academic Publishing, The Hague, Netherlands, pp. 45–93.

Allen, J. A. 1998. Mangroves as alien species: The case of Hawaii. Global Ecology and Biogeography Letters 7: 61–71.

Alongi, D. 2008. Mangrove forests: Resilience, protection from tsunamis, and responses to global climate change. Estuarine, Coastal and Shelf Science 76: 1–13.

Altor, A. E., and W. J. Mitsch. 2006. Methane flux from created wetlands: Relationship to intermittent versus continuous inundation and emergent macrophytes. Ecological Engineering 28: 224–234.

Altor, A. E., and W. J. Mitsch. 2008a. Methane emissions and carbon dioxide fluxes in created wetland mesocosms: Effects of hydrologic regime and hydric soils. Ecological Applications 18: 1307–1320.

Altor, A. E., and W. J. Mitsch. 2008b. Pulsing hydrology, methane emissions, and carbon dioxide fluxes in created marshes: A 2-year ecosystem study. Wetlands 28: 423–438.

Alvord, H. H., and R. H. Kadlec. 1996. Atrazine fate and transport in the Des Plaines Wetlands. Ecological Modelling 90: 97–107.

Amon, J. P., C. A. Thompson, Q. J. Carpenter, and J. Miner. 2002. Temperate zone fens of the glaciated Midwestern USA. Wetlands 22: 301–317.

Anderson, C. J., W. J. Mitsch, and R. W. Nairn. 2005. Temporal and spatial development of surface soil conditions in two created riverine marshes. Journal of Environmental Quality 34: 2072–2081.

Anderson, C. J., and W. J. Mitsch. 2005. Effect of pulsing on macrophyte productivity and nutrient uptake: A wetland mesocosm experiment. American Midland Naturalist 154: 305–319.

Anderson, C. J., and W. J. Mitsch. 2006. Sediment, carbon, and nutrient accumulation at two 10-year-old created riverine marshes. Wetlands 26: 779–792.

Anderson, C. J., and B. G. Lockaby. 2007. Soils and biogeochemistry of tidal freshwater forested wetlands. In W. H. Conner, T. W. Doyle, and K. W. Krauss, eds. *Ecology of Tidal Freshwater Forested Wetlands of the Southeastern United States*. Springer, Inc., Dordrecht, Netherlands, pp. 65–88.

Anderson, C. J., and W. J. Mitsch. 2008a. The influence of flood connectivity on bottomland forest productivity in central Ohio, USA. Ohio Journal of Science 108 (2): 2–8.

Anderson, C. J., and W. J. Mitsch. 2008b. Tree basal growth response to flooding in a bottomland hardwood forest in central Ohio. J. American Water Resources Association, 44: 1512–1520.

Anderson, D. C., and D. J. Cooper. 2000. Plant-herbivore-hydroperiod interactions: Effects of native mammals on floodplain tree recruitment. Ecological Applications 10: 1384–1399.

Anderson, F. O. 1976. Primary productivity in a shallow water lake with special reference to a reed swamp. Oikos 27: 243–250.

Anderson, R. C., and J. White. 1970. A cypress swamp outlier in southern Illinois. Illinois State Acad. Sci. Trans. 63: 6–13.

Antheunisse, A. M., and J. T. A. Verhoeven. 2008. Short-term responses of soil nutrient dynamics and herbaceous riverine plant communities to summer inundation. Wetlands 28: 232–244.

Antlfinger, A. E., and E. L. Dunn. 1979. Seasonal patterns of CO_2 and water vapor exchange of three salt-marsh succulents. Oecologia (Berl.) 43: 249–260.

Armentano, T. V., and E. S. Menges. 1996. Patterns of change in the carbon balance of organic soil wetlands of the temperate zone. Journal of Ecology 74: 755–774.

Aselmann, I., and P. J. Crutzen. 1989. Global distribution of natural freshwater wetlands and rice paddies, their net primary productivity, seasonality and possible methane emissions. Journal of Atmospheric Chemistry 8: 307–358.

Baldwin, A. H., K. L. McKee, and I. A. Mendelssohn. 1996. The influence of vegetation, salinity, and inundation on seed banks of oligohaline coastal marshes. American Journal of Botany 83: 470–479.

Baldwin, A. H., and I. A. Mendelssohn. 1998. Effects of salinity and water level on coastal marshes: An experimental test of disturbance as a catalyst for vegetation change. Aquatic Botany 61: 255–268.

Baldwin, A. H., M. S. Egnotovish, and E. Clarke. 2001. Hydrologic change and vegetation of tidal freshwater marshes: Field, greenhouse, and seed-bank experiments. Wetlands 21: 519–531.

Ball, M. C. 1980. Patterns of secondary succession in a mangrove forest in south Florida. Oecologia 44: 226–235.

Balogh, G. R., and T. A. Bookhout. 1989. Purple loosestrife (*Lythrum salicaria*) in Ohio's Lake Erie marshes. Ohio Journal of Science 89: 62–64.

Banaszuk, P., and A. Kamocki. 2008. Effects of climatic fluctuations and land-use changes on the hydrology of temperate fluviogenous mire. Ecological Engineering 32: 133–146.

Bartsch, I., and T. R. Moore. 1985. A preliminary investigation of primary production and decomposition in four peatlands near Schefferville, Quebéc. Canadian Journal of Botany 63: 1241–1248.

Batzer, D. P., and S. A. Wissinger. 1996. Ecology of insect communities in nontidal wetlands. Annual Review of Entomology 41: 75–100.

Bayley, S. E., and R. L. Mewhort. 2004. Plant community structure and functional differences between marshes and fens in the southern boreal region of Alberta, Canada. Wetlands 24: 277–294.

Bazilevich, N. I., and A. A. Tishkov. 1982. Conceptual balance model of chemical element cycles in a mesotrophic bog ecosystem. In *Ecosystem Dynamics in Freshwater Wetlands and Shallow Water Bodies*, vol. II. SCOPE and UNEP, Centre of International Projects, Moscow, USSR, pp. 236–272.

Beadle, L. C. 1974. *The Inland Waters of Tropical Africa*. Longman, London.

Beck, L. T. 1977. Distribution and Relative Abundance of Freshwater Macroinverebrates of the Lower Atchafalaya River Basin, Louisiana. Master's thesis, Louisiana State University, Baton Rouge.

Bedford, B. L., M. R. Walbridge, A. Aldous. 1999. Patterns of nutrient availability and plant diversity of temperate North American wetlands. Ecology 8: 1251–1269.

Bedford, B. L., and K. S. Godwin. 2003. Fens of the United States: Distribution, characteristics, and scientific connection versus legal isolation. Wetlands 23: 608–629.

Beier, C., and L. Rasmussen. 1994. Effects of whole-ecosystem manipulations on ecosystem internal processes. Trends in Ecology and Evolution 9: 218–223.

Bellamy, D. J., and J. Rieley. 1967. Some ecological statistics of a 'miniature bog.' Oikos 18: 33–40.

Bernard, J. M., and B. A. Solsky. 1977. Nutrient cycling in a *Carex lacustris* wetland. Canadian Journal of Botany 55: 630–638.

Bleuten, W., W. Borren, P.H. Blaser, T. Tsuchihara, E.D. Lapshina, M. Makila, D. Siegel, H. Joosten, and M.J. Wassen. 2006. Hydrological processes, nutrient flows and patterns of fens and bogs. In J.T.A. Verhoeven, B. Beltman, R. Bobbink, and D.F. Whigham, eds. *Wetlands and Natural Resource Management*. Springer, Heidelberg, pp. 183–204.

Botch, M. S., K. I. Kobak, T. S. Vinson, and T. P. Kolchugina. 1995. Carbon pools and accumulation in peatlands of the former Soviet Union. Global Biogeochemical Cycles 9: 37–46.

Bouchard, V., and J. C. Lefeuvre. 2000. Primary production and macro-detritus dynamics in a European salt marsh: Carbon and nitrogen budgets. Aquatic Botany 67: 23–42.

Bouchard, V., M. Tessier, F. Digaire, J. P. Vivier, L. Valery, J. C. Gloaguen, and J. C. Lefeuvre. 2003. Sheep grazing as management tool in Western European saltmarshes. C. R. Biologies 326: S148–S157.

Bouchard, V., S. D. Frey, J. M. Gilbert, and S. E. Reed. 2007. Effects of macrophytes functional group richness on emergent freshwater wetland functions. Ecology 88: 2903–2914.

Bouillon, S., A. V. Borges, E. Castañeda-Moya, K. Diele, T. Dittmar, N. C. Duke, E. Kristensen, S. Y. Lee, C. Marchand, J. J. Middelburg, V. H. Rivera-Monroy, T. J. Smith III, and R. R. Twilley. 2008. Mangrove production and carbon sinks: A revision of global budget estimates. Global Biogeochemistry Cycles 22: GB2013.

Bowden, W. B., C. J. Vorosmarty, J. T. Morris, B. J. Peterson, J. E. Hobbie, P. A. Steudler, and B. Moore. 1991. Transport and processing of nitrogen in a tidal freshwater wetland. Water Resources Research 27: 389–408.

Boyd, C. E. 1971. The dynamics of dry matter and chemical substances in a Juncus efusus population. American Midland Naturalist 86: 28–45.

Bradley, P. M., and E. L. Dunn. 1989. Effects of sulfide on the growth of three salt marsh halophytes of the southeastern United States. American Journal of Botany 76: 1707–1713.

Bridgham, S. D., J. Pastor, J. A. Janssens, C. Chapin, and T. J. Malterer. 1996. Multiple limiting gradients in peatlands: A call for a new paradigm. Wetlands 16: 45–65.

Bridgham, S. D., K. Updegraff, and J. Pastor. 1998. Carbon, nitrogen, and phosphorus mineralization in northern wetlands. Ecology 79: 1545–1561.

Bridgham, S. D., J. Pastor, K. Updegraff, T. J. Malterer, K. Johnson, C. Harth, and J. Chen. 1999. Ecosystem control over temperature and energy flux in northern peatlands. Ecological Applications 9: 1345–1358.

Bridgham, S. D., K. Updegraff, and J. Pastor. 2001. A comparison of nutrient availability indices along an ombrotrophic-minerotrophic gradient in Minnesota wetlands. Soil Science Society of America Journal 65: 259–269.

Brinson, M. M. 1977. Decomposition and nutrient exchange of litter in an alluvial swamp forest. Ecology 58: 601–609.

Brinson, M. M. 1993. A Hydrogeomorphic Classification for Wetlands. Wetlands Research Program Technical Report WRP-DE-4. U.S. Army Engineer Waterways Experiment Station, Vicksburg, MS.

Brinson, M. M., B. L. Swift R. C. Plantico, and J. S. Barclay. 1981. Riparian Ecosystems: Their Ecology and Status, U.S. Fish and Wildlife Service, Biol. Serv. Prog., FWS/OBS-81/17, Washington, DC, 151 pp.

Brix, H., B. K. Sorrell, and B. Lorenzen. 2001. Are *Phragmites*-dominated wetlands a net source or net sink of greenhouse gases? Aquatic Botany 69: 313–324.

Brody, M., and E. Pendleton. 1987. FORFLO: A Model to Predict Changes in Bottomland Hardwood Forests. Office of Information Transfer, U.S. Fish & Wildlife Service, Fort Collins, CO.

Brown, C. A. 1984. Morphology and biology of cypress trees. In K. C. Ewel and H. T. Odum, eds. *Cypress Swamps*. Univ. Presses of Florida, Gainesville. pp. 16–24.

Brown, M. T. 1988. A simulation model of hydrology and nutrient dynamics in wetlands. Computers, Environment and Urban Systems 12(4): 221–237.

Brown, S. L. 1978. A Comparison of Cypress Ecosystems in the Landscape of Florida. Ph.D. dissertation, University of Florida, 569 pp.

Brown, S. L. 1981. A comparison of the structure, primary productivity, and transpiration of cypress ecosystems in Florida. Ecological Monographs 51: 403–427.

Brown, S. L., and A. E. Lugo. 1982. A comparison of structural and functional characteristics of saltwater and freshwater forested wetlands. In B. Gopat, R. E. Turner, R. G. Wetzel, and D. F. Whigham, eds. *Wetlands: Ecology and Management*. National Institute of Ecology and International Scientific Publications, Jaipur, India, pp. 109–130.

Brown, S. L., and D. L. Peterson. 1983. Structural characteristics and biomass production of two Illinois bottomland forests. American Midland Naturalist 110: 107–117.

Brueske, C. C., and G. W. Barrett. 1994. Effects of vegetation and hydrologic load on sedimentation patterns in experimental wetland ecosystems. Ecological Engineering 3: 429–447.

Bryan, C. F., D. J. DeMoss, D. S. Sabins, and J. PP. Newman, Jr. 1976. *A Limnological Survey of the Atchafalaya Basin*, Annual Report, Louisiana Cooperative Fishery Research Unit, School of Forestry and Wildlife Management, Louisiana State University, Baton Rouge, 285 pp.

Bubier, J. L., G. Bhatia, T. R. Moore, N. T. Roulet, and P. M. Lafleur. 2003. Spatial and temporal variability in growing-season net ecosystem carbon dioxide exchange at a large peatland in Ontario, Canada. Ecosystems 6: 353–367.

Bubier, J., T. Moore, K. Savage, and P. Crill. 2005. A comparison of methane flux in a boreal landscape between a dry and a wet year. Global Biogeochemical Cycles 19: GB1023.

Busnardo, M. J., R. M. Gersberg, R. Langis, T. L. Sinicrope, and J. B. Zedler. 1992. Nitrogen and phosphorus removal by wetland mesocosms subjected to different hydroperiods. Ecological Engineering 1: 287–307.

Buxton, R. P., P. N. Johnson, and P. R. Espie. 1996. Sphagnum Research Programme: The Ecological Effects of Commercial Harvesting. Science for Conservation: 25. New Zealand Department of Conservation, Wellington, 33 pp.

Caldwell, R., and Crow. 1992. A floristic and vegetation analysis of a tidal freshwater marsh on the Merrimac River, West Newberry, MA. Rhodora 94: 63–97.

Campeau, S., H. R. Murkin, and R. D. Titman. 1994. Relative importance of algae and emergent plant litter to freshwater marsh invertebrates. Canadian Journal of Fisheries and Aquatic Sciences 51: 681–692.

Carpenter, S. R., J. F. Kitchell, K. L. Cottingham, D. E. Schindler, D. L. Christensen, D. M. Post, and N. Voichick. 1996. Chlorophyll variability, nutrient input and grazing: evidence from whole-lake experiments. Ecology 77: 725–735.

Carpenter, S. R., J. J. Cole, J. F. Kitchell, and M. L. Pace. 1998. Impact of dissolved organic carbon, phosphorus and grazing on phytoplankton biomass and production in experimental lakes. Limnology & Oceanography 43: 73–80.

Carter, M. R., L. A. Bums, T. R. Cavinder, K. R. Dugger, P. L. Fore, D. B. Hicks, H. L. Revells, and T. W. Schmidt. 1973. Ecosystem Analysis of the Big Cypress Swamp and Estuaries, U.S. EPA 904/9-74[sim]]02, Region IV, Atlanta.

Casey, W. P., and K. C. Ewel. 2006. Patterns of succession in forested depressional wetlands in North Florida, USA. Wetlands 26: 147–160.

Cameron, C. C. 1970. Peat deposits of northeastern Pennsylvania. U.S. Geological Survey Bulletin 1317-A. 90 pp.

Chalmers, A. G., R. G. Wiegert, and P. L. Wolff. 1985. Carbon balance in a salt marsh: Interactions of diffusive export, tidal deposition and rainfall-caused erosion. Estuarine Coastal and Shelf Science 21: 757–771.

Chapin, C. T., and J. Pastor. 1995. Nutrient limitations in the northern pitcher plant *Sarracenia purpurea*. Canadian Journal of Botany 73: 728–734.

Chapin, C. T., S. D. Bridgham, and J. Pastor. 2004. pH and nutrient effects on above-ground net primary production in a Minnesota, USA bog and fen. Wetlands 24: 186–201.

Chapman, S. B. 1965. The ecology of Coom Rigg Moss, Northumberland: Some water relations of the bog system. Journal of Ecology 53: 371–384.

Chapman, V. J. 1960. *Salt Marshes and Salt Deserts of the World*, Interscience, New York. 392 pp.

Chapman, V. J. 1976a. *Coastal Vegetation*, 2nd ed., Pergamon Press, Oxford, UK. 292 pp.

Chapman, V. J. 1976b. *Mangrove Vegetation*, J. Cramer, Vaduz, Germany. 447 pp.

Chapman, V. J. 1977. *Wet Coastal Ecosystems*, Elsevier, Amsterdam, 428 pp.

Chen, R., and R. R. Twilley. 1998. A gap dynamic model of mangrove forest development along gradients of soil salinity and nutrient resources. Journal of Ecology 86: 37–51.

Childers, D. L. 1994. Fifteen years of marsh flumes: a review of marsh-water column interactions in southeastern USA estuaries. In W. J. Mitsch, ed. *Global Wetlands: Old World and New*. Elsevier, Amsterdam, pp. 277–293.

Childers, D. L., J. W. Day, Jr., and H. N. McKellar, Jr. 2000. Twenty more years of marsh and estuarine flux studies: revisiting Nixon (1980). In M. P. Weinstein and D. A. Kreeger, eds. *Concepts and Controversies in Tidal Marsh Ecology*. Kluwer Academic Publishers, Netherlands, pp. 389–421.

Childers, D. L., R. F. Doren, R. Jones, G. B. Noe, M. Rugge, and L. J. Scinto. 2003. Decadal change in vegetation and soil phosphorus pattern across the Everglades landscape. Journal of Environmental Quality 32: 344–362.

Chimney, M. J., and G. Goforth. 2006. History and description of the Everglades Nutrient Removal Process, a subtropical constructed wetland in southern Florida (USA). Ecological Engineering 27: 268–278.

Christensen, N., W. J. Mitsch, and S. E. Jørgensen. 1994. A first generation ecosystem model of the Des Plaines River experimental wetlands. Ecological Engineering 3: 495–521.

Cintrón, G., A. E. Lugo, and R. Martinez. 1985. Structural and functional properties of mangrove forests. In W. G. D'Arcy and M. D. Corma, eds. *The Botany and Natural History of Panama*. Missouri Botanical Garden, St. Louis, pp. 53–66.

Cintrón, G., A. E. Lugo, D. J. Pool, and G. Morris. 1978. Mangroves of arid environments in Puerto Rico and adjacent islands. Biotropica 10: 110–121.

Clarkson, B. R., K. Thompson, L. A. Schipper, and M. McLeod. 1999. Moanatuatua Bob—proposed restoration of a New Zealand restiad peat bog ecosystem. In W. Streever, ed. *An International Perspective on Wetland Rehabilitation*. Kluwer Academic Publishers, Dordrecht, The Netherlands, pp. 127–137.

Clewell, A. F., and D. B. Ward. 1987. White cedar in Florida and along the northern Gulf Coast. In A.D. Laderman, ed. *Atlantic White Cedar Wetlands*. Westview Press, Boulder, CO. pp. 69–82.

Clymo, R. S. 1970. The growth of *Sphagnum*: Methods of measurement. Journal of Ecology 58: 13–49.

Clymo, R. S., and E. J. F. Reddaway. 1971. Productivity of *Sphagnum* (bog-moss) and peat accumulation. Hydrobiologia (Bucharest) 12: 181–192.

Clymo, R S., and P. M. Hayward. 1982. The ecology of *Sphagnum*. In A. J. E. Smith, ed. *Bryophyte Ecology*. Chapman and Hall, London, pp. 229–289.

Cochran, M. 2001. Effect of hydrology on bottomland hardwood forest productivity in central Ohio (USA). Master's thesis, Natural Resources, The Ohio State University, Columbus.

Cole, T. G., K. C. Ewel, and N. N. Devoe. 1999. Structure of mangrove trees and forests in Micronesia. Forest Ecology and Management 117: 95–109.

Coleman, J. M., and H. H. Roberts. 1989. Deltaic coastal wetlands. Geologie en Mijnbouw 68: 1–24.

Coles, B., and J. Coles. 1989. *People of the Wetlands, Bogs, Bodies and Lake-dwellers*. Thames and Hudson, New York, NY, 215 pp.

Conner, W. H., and J. W. Day, Jr. 1976. Productivity and composition of a bald cypress-water tupelo site and a bottomland hardwood site in a Louisiana swamp. American Journal of Botany 63: 1354–1364.

Conner, W. H., J. G. Gosselink, and R. T. Parrondo. 1981. Comparison of the vegetation of three Louisiana swamp sites with different flooding regimes. American Journal of Botany 68: 32–331.

Conner, W. H., and J. W. Day, Jr. 1982. The ecology of forested wetlands in the southeastern United States. In B. Gopal, R. E. Turner, R. G. Wetzel, and D. F. Whigham, eds. *Wetlands: Ecology and Management*. National Institute of Ecology and International Scientific Publications, Jaipur, India, pp. 69–87.

Conner, W. H., I. Mihalia, and J. Wolfe. 2002. Tree community structure and changes from 1987 to 1999 in three Louisiana and three South Carolina forested wetlands. Wetlands 22: 58–70.

Costanza, R., and A. Voinov, eds. 2004. *Landscape Simulation Modeling: A Spatially Explicit, Dynamic Approach*. Springer, Secaucus, New Jersey, 330 pp.

Costanza, R., W. J. Mitsch, and J. W. Day. 2006. A new vision for New Orleans and the Mississippi Delta: Applying ecological economics and ecological engineering. Frontiers in Ecology and the Environment 4: 465–472.

Cowardin, L. M., V. Carter, F. C. Golet, and E. T. LaRoe. 1979. Classification of Wetlands and Deepwater Habitats of the United States, U.S. Fish & Wildlife Service Pub. FWS/OBS-79/31, Washington, DC, 103 pp.

Craft, C. B., J. Vymazal, and C. J. Richardson. 1995. Response of Everglades plant communities to nitrogen and phosphorus additions. Wetlands 15: 258–271.

Craft, C., J. Clough, J. Ehman, S. Joye, D. Park, S. Pennings, H. Guo, and M. Machmuller. 2009. Forecasting the effects of accelerated sea-level rise on tidal marsh ecosystem services. Frontiers in Ecology and the Environment. in press

Cragg, J. B. 1961. Some aspects of the ecology of moorland animals. J. Anim Ecol. 30: 205–234.

Crain, C. M., B. R. Silliman, S. L. Bertness, and M. D. Bertness. 2004. Physical and biotic drivers of plant distribution across estuarine salinity gradients. Ecology 85: 2539–2549.

Cromarty, D., and D. Scott. 1996. *A Directory of Wetlands in New Zealand*. Department of Conservation, Wellington, New Zealand.

Cronk, J. K., and W. J. Mitsch. 1994a. Aquatic metabolism in four newly constructed freshwater wetlands with different hydrologic inputs. Ecological Engineering 3: 449–468.

Cronk, J. K., and W. J. Mitsch. 1994b. Periphyton productivity on artificial and natural surfaces in four constructed freshwater wetlands under different hydrologic regimes. Aquatic Botany 48: 325–341.

Crozier, C. R., and R. D. DeLaune. 1996. Methane production by soils from different Louisiana marsh vegetation types. Wetlands 16: 121–126.

Dabel, C. V., and F. P. Day, Jr. 1977. Structural composition of four plant communities in the Great Dismal Swamp, Virginia. Torrey Botanical Club Bulletin 104: 352–360.

Dahl, T. E. 2006. *Status and Trends of Wetlands in the Conterminous United States 1998 to 2004*. U.S. Department of the Interior, Fish and Wildlife Service, Washington, DC, 112 pp.

Dahl, T. E., and C. E. Johnson. 1991. *Wetlands Status and Trends in the Conterminous United States Mid-1970s to mid-1980s*. U.S. Department of Interior, Fish and Wildlife Service, Washington, DC, 28 pp.

Damman, A. W. H. 1978. Distribution and movement of elements in ombrotrophic peat bogs. Oikos 30: 480–495.

Damman, A. W. H. 1979. Geographic Patterns in Peatland Development in Eastern North America. In Proceedings of the International Symposium of Classification of Peat and Peatlands, Hyytiala, Finland, International Peat Society, pp. 42–57.

Damman, A.W.H. 1986. Hydrology, development, and biogeochemistry of ombrogenous peat bogs with special reference to nutrient relocation in a western Newfoundland bog. Canadian Journal of Botany 64: 384–391.

Damman, A. W. H., and T. W. French. 1987. The Ecology of Peat Bogs of the Glaciated Northeastern United States: a Community Profile. U.S. Fish and Wildlife Service, Washington, DC. 100 pp.

Danielsen, F., M. K. Sørensen, M. F. Olwig, V. Selvam, F. Parish, N. D. Burgess, T. Hiraishi, V. M. Karunagaran, M. S. Rasmussen, L. B. Hansen, A. Quarto, and N. Suryadiputra. 2005. The Asian tsunami: A protective role for coastal vegetation. Science 310: 643.

Davis, J. H. 1940. The ecology and geologic role of mangroves in Florida. Carnegie Institution, Washington. Publ. No. 517, pp. 303–412.

Day, F. P. 1984. Biomass and litter accumulation in the Great Dismal Swamp. In K.C. Ewel and H.T. Odum, eds. *Cypress Swamps*, University of Florida Press, Gainesville, pp. 386–392.

Day, J. W., Jr., T. J. Butler, and W. G. Conner. 1977. Productivity and nutrient export studies in a cypress swamp and lake system in Louisiana. In M. Wiley, ed. *Estuarine Processes*, Vol. II, Academic Press, New York, pp. 255–269.

Day, J. W., Jr., C. Coronado-Molina, F. R. Vera-Herrera, R. Twilley, V. H. Rivera-Monroy, H. Alvarez-Guillen, R. Day, and W. Conner. 1996. A 7 year record of above-ground net primary production in an southeastern Mexican mangrove forest. Aquatic Botany 55: 39–60.

Day, J. W., J. Ko, J. Rybczyk, D. Sabins, R. Bean, G. Berthelot, C. Brantley, L. Cardoch, W. Conner, J.N. Day, A.J. Englande, S. Feagley, E. Hyfield, R. Lane, J. Lindsey, J. Mistich, E. Reyes, and R. Twilley. 2004. The use of wetlands in the Mississippi delta for wastewater assimilation: a review. Ocean & Coastal Management 47: 671–691.

Day J. W., Jr., D. F. Boesch, E. J. Clairain, G. P. Kemp, S. B. Laska, W. J. Mitsch, K. Orth, H. Mashriqui, D. R. Reed, L. Shabman, C. A. Simenstad, B. J. Streever, R. R. Twilley, C. C. Watson, J. T. Wells, and D. F. Whigham. 2007. Restoration of the Mississippi Delta: Lessons From Hurricanes Katrina and Rita. Science 315: 1679–1684.

Day, R. T., P. A. Keddy, J. M. McNeil, and T. J. Carleton. 1988. Fertility and disturbance gradients: a summary model for riverine marsh vegetation. Ecology 69: 1044–1054.

DeGraaf, R. M., and D. D. Rudis. 1986. New England Wildlife: Habitat Natural History, and Disturbance. General Technical Report NE-108, Northeastern Forest Experiment Station, Forest Service, U.S. Department of Agriculture, Washington, DC, 491 pp.

DeGraaf, R. M., and D. D. Rudis. 1990. Herpetofaunal species composition and relative abundance among three New England forest types. Forest Ecology and Management 32: 155–165.

Delaune, R. D., and C. W. Lindau. 1990. Fate of added 15N labelled nitrogen in a *Sagittaria lancifolia* L. Gulf Coast marsh. Journal of Freshwater Ecology 5: 429–431.

Denny, P. 1985. Wetland plants and associated plant life-forms. In P. Denny, ed. *The Ecology and Management of Wetland Vegetation*. Junk, Dordrecht, pp. 1–18.

Denny, P. 1993. Wetlands of Africa: Introduction. Pages 1–31 In D. F. Whigham, D. Dykyjová and S. Hejny, eds. *Wetlands of the World I. Inventory, Ecology and Management*. Kluwer, Dordrecht, The Netherlands.

Dent, D. L. 1986. Acid Sulphate Soils: A Baseline for Research and Development. ILRI Publication 39, Wageninggen, Netherlands.

Dent, D. L. 1992. Reclamation of acid sulphate soils. In R. Lal and B. A. Stewart, eds. *Soil Restoration, Advances in Soil Science*, Vol. 17. Springer-Verlag, New York, pp. 79–122.

Derksen, A. J. 1989. Autumn movements of underyearling northern pike, *Esox lucius*, from a large Manitoba marsh. Canadian Field Naturalist 103: 429–431.

DeRoia, D. M., and T. A. Bookhout. 1989. Spring feeding ecology of teal on the Lake Erie marshes (abstract). Ohio Journal of Science 89(2): 3.

deSzalay, F. A., and V. H. Resh. 1996. Spatial and temporal variability of trophic relationships among aquatic macrophytes in a seasonal marsh. Wetlands 16: 458–466.

deSzalay, F. A., and V. H. Resh. 1997. Responses of wetland invertebrates and plants important in waterfowl diets to burning and mowing of emergent vegetation. Wetlands 17: 149–156.

de la Cruz, A. A. 1974. Primary productivity of coastal marshes in Mississippi. Gulf Research Reports 4: 351–356.

Dierberg, F. E., and R. L. Brezonik. 1983a. Tertiary treatment of municipal wastewater by cypress domes. Water Research 17: 1027–1040.

Dierberg, F. E., and R. L. Brezonik. 1983b. Nitrogen and phosphorus mass balances in natural and sewage-enriched cypress domes. Journal of Applied Ecology 20: 323–337.

Dierberg, F. E., and P. L. Brezonik. 1985. Nitrogen and phosphorus removal by cypress swamp sediments. Water, Air and Soil Pollution 24: 207–213.

Dorge, C. L., W. J. Mitsch, and J. R. Wiemhoff. 1984. Cypress wetlands in southern Illinois. In K C. Ewel and H. T. Odum, eds. *Cypress Swamps*, University Presses of Florida, Gainesville, pp. 393–404.

Doyle, G. J. 1973. Primary production estimates of native blanket bog and meadow vegetation growing on reclaimed peat at Glenamoy, Ireland. In L. C. Bass and F. E. Wielgolaski, eds. *Primary Production and Production Processes, Tundra Biome*, Tundra Biome Steering Committee, Edmonton and Stockholm, pp. 141–151.

Drexler, J. Z., and B. L. Bedford. 2002. Pathways of nutrient loading and impacts on plant diversity in a New York peatland. Wetlands 22: 263–281.

Dreyer, G. D., and W. A. Niering, eds. 1995. Tidal Marshes of Long Island Sound: Ecology, History and Restoration. The Connecticut College Arboretum, Bulletin No. 34, New London, CT.

Dudek, D. M., J. R. McClenahen, and W. J. Mitsch. 1998. Tree growth responses of *Populus deltoides* and *Juglans nigra* to streamflow and climate in a bottomland hardwood forest in Central Ohio. American Midlland Naturalist 140: 233–244.

Duever, M. J., J. E. Carlson, and L. A. Riopelle. 1984. Corkscrew Swamp: a virgin cypress strand. In K. C. Ewel and H. T. Odum, eds. *Cypress Swamps*. University Presses of Florida, Gainesville, pp. 334–348.

Du Rietz, G. E. 1949. Huvudenheter och huvudgränser i Svensk myrvegetation. Svensk Botanisk Tidkrift 43: 274–309.

Dvorak, J. 1978. Macrofauna of invertebrates in helophyte communities. In D. Dykyjova and J. Kvet eds. Pond Littoral Ecosystems. Springer-Verlag, Berlin, pp. 389–392.

Dykyjova, D., and J. Kvet, eds. 1978. *Pond Littoral Ecosystems*. Springer-Verlag, Berlin, 464 pp.

Effler, R. S., G. P. Shaffer, S. S. Hoeppner, and R. A. Goyer. 2007. Ecology of the Maurepas Swamp: Effects of salinity, nutrients, and insect defoliation. Soils and biogeochemistry of tidal freshwater forested wetlands. In W.H. Conner, T.W. Doyle, and K.W. Krauss, eds. *Ecology of Tidal Freshwater Forested Wetlands of the Southeastern United States*. Springer, Dordrecht, Netherlands, pp. 349–384.

Egler, F. E. 1952. Southeast saline Everglades vegetation, Florida, and its management. Vegetatio Acta Geobotica 3: 213–265.

Eisenlohr, W. S. 1972. Hydrologic Investigations of Prairie Potholes in North Dakota, 1958–1969. Professional Paper 585, U.S. Geological Survey, Washington, DC.

Evers, D. E., C. E. Sasser, J. G. Gosselink, D. A. Fuller, and J. M. Visser. 1998. The impact of vertebrate herbivores on wetland vegetation in Atchafalaya Bay, Louisiana. Estuaries 21: 1–13.

Ewel, K. C., and W. J. Mitsch. 1978. The effects of fire on species composition in cypress dome ecosystems. Florida Scientist 41: 25–31.

Ewel, K. C., and H. T. Odum, eds. 1984. *Cypress Swamps*. University Presses of Florida, Gainesville, FL. 472 pp.

Ewel, K C., and L. P. Wickenheiser. 1988. Effects of swamp size on growth rates of cypress (*Taxodium distichum*) trees. American Midland Naturalist 120: 362–370.

Ewel, K. C., R. R. Twilley, and J. E. Ong. 1998. Different kinds of mangrove forests provide different goods and services. Global Ecology and Biogeography Letters 7: 83–94.

Farrar, J., and R. Gersib. 1991. Nebraska Salt Marshes: Last of the Least. Nebraska Game and Park Commission, Lincoln, NE, 23 pp.

Feller, I. C., C. E. Lovelock, and K. L. McKee. 2007. Nutrient addition differentially affects ecological processes of *Avicennia germinans* in nitrogen versus phosphorus limited mangrove ecosystems. Ecosystems 10: 347–359.

Feminella, J. W., and V. H. Resh. 1989. Submersed macrophytes and grazing crayfish: an experimental study of herbivory in a California freshwater marsh. Holarctic Ecology 12: 1–8.

Fennessy, M.S., C. Brueske, and W.J. Mitsch. 1994a. Sediment deposition patterns in restored freshwater wetlands using sediment traps. Ecological Engineering 3: 409–428.

Fennessy, M. S., J. K. Cronk, and W. J. Mitsch. 1994b. Macrophyte productivity and community development in created freshwater wetlands under experimental hydrologic conditions. Ecological Engineering 3: 469–484.

Fennessy, M. S., and W. J. Mitsch. 2001. Effects of hydrology on spatial patterns of soil development in created riparian wetlands. Wetlands Ecology and Management 9: 103–120.

Field, D. W., A. Reyer, P. Genovese, and B. Shearer. 1991. Coastal wetlands of the United States—An accounting of a valuable national resource. Strategic Assessment Branch, Ocean Assessments Division, Office of Oceanography and Marine Assessments, National Ocean Service, National Oceanic and Atmosphere Administration, Rockville, MD.

Fink, D. F., and W. J. Mitsch. 2007. Hydrology and biogeochemistry in a created river diversion oxbow wetland. Ecological Engineering 30: 93–102.

Fitz, C., F. Sklar, T. Waring, A. Voinov, R. Costanza, and T. Maxwell. 2003. Development and application of the Everglades landscape model. In R.

Costanza and A. Voinov, eds. *Landscape Simulation Modeling: A Spatially Explicit, Dynamic Approach*. Springer, New York, pp. 143–172.

Fitz, C., N. Wang, J. Godin, A. Morales, Y. Wu, and F. Sklar. 2006. Application of the Everglades Landscape Model for Restoration Initiatives. Everglades Division, South Florida Water Management District, unpublished presentation.

Forrest, G. I., and R. A. H. Smith. 1975. The productivity of a range of blanket bog types in the Northern Pennines. Journal of Ecology 63: 173–202.

Fowler, K. 1987. Primary production and temporal variation in the macrophyte community of a tidal freshwater swamp. M.A. Thesis, Virginia Institute of Marine Science, College of William and Mary, Williamsburg, Virginia.

Frayer, W. E., T. J. Monahan, D. C. Bowden, and F. A. Graybill. 1983. Status and Trends of Wetlands and Deepwater Habitat in the Conterminous United States, 1950s to 1970s, Dept. of Forest and Wood Sciences, Colorado State University, Fort Collins, 32 pp.

Fuller, D. A., G. W. Peterson, R. K. Abernethy, and M. A. LeBlanc. 1988. The distribution and habitat use of waterfowl in Atchafalaya Bay, Louisiana. In C. E. Sasser and D. A. Fuller, eds. *Vegetation and Waterfowl Use of Islands in Atchafalaya Bay*. Coastal Ecology Institute, Louisiana State University, Baton Rouge, LA, pp. 73–103.

Funk, D. W., L. E. Noel, and A. H. Freedman. 2004. Environmental gradients, plant distribution, and species richness in artic salt marsh near Prudhoe Bay, Alaska. Wetlands Ecology and Management 12: 215–233.

Galatowitsch, S. M., N. O. Anderson, and P.D. Ascher. 1999. Invasiveness in wetland plants in temperate North America. Wetlands 19: 733–755.

Gallagher, J. L., and F. C. Daiber. 1974. Primary production of edaphic algae communities in a Delaware salt marsh. Limnology and Oceanography 19: 390–395.

Gerritsen, J., and H. S. Greening. 1989. Marsh seed banks of the Okefenokee Swamp: effects of hydrologic regime and nutrients. Ecology 70: 750–763.

Glaser, P. H. 1987. The Ecology of Patterned Boreal Peatlands of Northern Minnesota: A Community Prof. U.S. Fish Wildlife Service Report 85 (7.14), Washington, DC, 98 pp.

Glaser, P. H., D. I. Siegel, E. A. Romanowicz, and Y. P. Shen. 1997a. Regional linkages between raised bogs and the climate, groundwater, and landscape of north-western Minnesota. Journal of Ecology 85: 3–16.

Glaser, P. H., P. C. Bennett, D. I. Siegel, and E. A. Romanowicz. 1997b. Palaeo-reversals in groundwater flow and peatland development at Lost River, Minnesota, USA. Holocene 6: 413–421.

Glob, P. V. 1969. *The Bog People: Iron Age Man Preserved*, trans. by R.Bruce-Mitford, Cornell University Press, Ithaca, New York, 200 pp.

Golet, F. C., and D. J. Lowry. 1987. Water regimes and tree growth in Rhode Island Atlantic white cedar swamps. In A.D. Laderman, ed. *Atlantic White Cedar Wetlands*, Westview Press, Boulder, CO. pp. 91–110.

Golet, F. C., A. J. K. Calhoun, W. R. DeRagon, D. J. Lowry, and A. J. Gold. 1993. Ecology of Red Maple Swamps in the Glaciated Northeast: A Community Profile. Biological Report 12, U.S. Fish and Wildlife Service, Washington, DC, 151 pp.

Golley, F. B., H. T. Odum, and R. F. Wilson. 1962. The structure and metabolism of a Puerto Rican red mangrove forest in May. Ecology 43: 9–19.

Gomez, M. M., and F. P. Day, Jr. 1982. Litter nutrient content and production in the Great Dismal Swamp. American Journal of Botany 69: 1314–1321.

Gorham, E., 1967. Some Chemical Aspects of Wetland Ecology, Technical Mem. Committee on Geotechnical Research, National Research Council of Canada, No. 90, pp. 2–38.

Gorham, E. 1974. The relationship between standing crop in sedge meadows and summer temperature. Journal of Ecology 62: 487–491.

Gorham, E. 1991. Northern peatlands: role in the carbon cycle and probable responses to climatic Warming. Ecological Applications 1: 182–195.

Gorham, E., and J. A. Janssens. 1992. Concepts of fen and bog reexamined in relation to bryophyte cover and the acidity of surface waters. Acta Societatis Botanicorum Poloniae 61: 7–20.

Gorham, E., S. J. Eisenreich, J. Ford, and M. V. Santelmann. 1984. The chemistry of bog waters. In: W. Strum, ed. Chemical Processes in Lakes. John Wiley and Sons, Inc., New York, pp. 339–363.

Gosselink, J. G., Coleman, J. M., and R. E. Stewart, Jr. 1998. Coastal Louisiana. In M. J. Mac, P. A. Opler, C. E. Puckett Haecker, and P. D. Doran. eds. *Status and Trends of the Nation's Biological Resources*. Vol. 1. U.S. Department of the Interior, U.S. Geological Survey, Reston, VA, pp. 385–436.

Grace, J. B., and M. A. Ford. 1996. The potential impact of herbivores on the susceptibility of the marsh plant *Sagittaria lancifolia* to saltwater intrusion in coastal wetlands. Estuaries 19: 13–20.

Greenberg, R., J. E. Maldonado, S. Droege, and M. V. McDonald. 2006. Tidal marshes: A global perspective on the evolution and conservation of their terrestrial vertebrates. BioScience 56: 675–685.

Gribsholt, B., and E. Kristensen. 2002. Effects of bioturbation and plant roots on salt marsh biogeochemistry: A mesocosm study. Mar. Ecol. Prog. Ser. 241: 71–87.

Grosse, W., H. B. Büchel, and S. Lattermann. 1998. Root aeration in wetland trees and its ecophysiological significance. In A. D. Laderman, ed. *Coastally Restricted Forests*. Oxford University Press, New York, NY, pp. 293–305.

Hackney, C. T., G. B. Avery, L. A. Leonard, M. Posey, and T. Alphin. 2007. Biological, chemical, and physical characteristics of tidal freshwater swamp forests of the lower Cape Fear River/Estuary, North Carolina. In W. H. Conner, T. W. Doyle, and K. W. Krauss, eds. *Ecology of Tidal Freshwater Forested Wetlands of the Southeastern United States*. Springer, Inc., Dordrecht, Netherlands. pp. 183–221.

Hall, J. V., W. E. Frayer, and B. O. Wilen. 1994. Status of Alaska Wetlands. U.S. Fish & Wildlife Service, Alaska Region, Anchorage, AK, 32 pp.

Hall, S. L., and F. M. Fisher, Jr. 1985. Annual productivity and extracellular release of dissolved organic compounds by the epibenthic algal community of a brackish marsh. Journal of Phycology 21: 277–281.

Harris, L. D. 1989. The faunal significance of fragmentation of southeastern bottomland forests. In D. D. Hook and R. Lea, eds. *Proceedings of the Symposium the Forested Wetlands of the Southern United States*, U.S. Department of Agriculture, Forest Service, Southeastern Forest Experiment Station, Asheville, NC, pp. 126–134.

Harter, S. K., and W. J. Mitsch. 2003. Patterns of short-term sedimentation in a freshwater created marsh. Journal of Environmental Quality 32: 325–334.

Hatton, R. S. 1981. Aspects of Marsh Accretion and Geochemistry: Barataria Basin, La., Master's thesis, Louisiana State University, Baton Rouge, LA.

Heal, O., W., H. E. Jones, and J. B. Whittaker. 1975. Moore House, U.K. In: T. Rosswall and O. W. Heal, eds. *Structure and Function of Tundra Ecosystems*. Ecological Bulletin 20, Swedish Natural Science Research Council, Stockholm, pp. 295–320.

Heald E. J. 1971. The Production of Organic Detritus in a South Florida Estuary, University of Miami Sea Grant Technical Bulletin No. 6, Coral Gables, FL, 110 pp.

Heilman, P. E. 1968. Relationship of availability of phosphorus and cations to forest succession and bog formation in interior Alaska. Ecology 49: 331–336.

Heinselman, M. L. 1970. *Landscape evolution and peatland types, and the Lake Agassiz Peatlands Natural Area, Minnesota*. Ecological Monographs 40: 235–261.

Hemond, H. F. 1980. *Biogeochemistry of Thoreau's Bog, Concord, Mass*. Ecological Monographs 50: 507–526.

Hemond, H. F. 1983. The nitrogen budget of Thoreau's Bog. Ecology 64. 99–109.

Hernandez, M. E., and W. J. Mitsch. 2006. Influence of hydrologic pulses, flooding frequency, and vegetation on nitrous oxide emissions from created riparian marshes. Wetlands 26: 862–877.

Hernandez, M. E., and W. J. Mitsch. 2007a. Denitrification in created riverine wetlands: Influence of hydrology and season. Ecological Engineering 30: 78–88.

Hernandez, M. E., and W. J. Mitsch. 2007b. Denitrification potential and organic matter as affected by vegetation community, wetland age, and plant introduction in created wetlands. Journal of Environmental Quality 36: 333–342.

Hey, D. L., M. A. Cardamone, J. H. Sather, and W. J. Mitsch. 1989. Restoration of riverine wetlands: the Des Plaines river wetlands demonstration project. In W. J. Mitsch and S. E. Jørgensen, eds. *Ecological Engineering: An Introduction to Ecotechnology*, John Wiley, New York, pp. 159–183.

Hey, D. L., K. R. Barrett, and C. Biegen. 1994a. The hydrology of four experimental constructed marshes. Ecological Engineering 3: 319–343.

Hey, D. L., A. L. Kenimer, and K. R. Barrett. 1994b. Water quality improvement by four experimental wetlands. Ecological Engineering 3: 381–398.

Holm, G. O., Jr. 1998. Comparisons of the Atchafalaya and Wax Lake Deltas: Salinity incursions and the loss of *Sagittaria latifolia* Willd. M.S. Thesis. Louisiana State University, Baton Rouge, Louisiana.

Hopkinson, C. S. 1985. Shallow water benthic and pelagic metabolism: Evidence for heterotrophy in the nearshore. Marine Biology 87: 19–32.

Hopkinson, C. S., Jr., J. G. Gosselink, and F. T. Parrondo. 1980. Production of coastal Louisiana marsh plants calculated from phenometric techniques. Ecology 61: 1091–1098.

Hornung, J. P., and L. Foote. 2006. Aquatic invertebrate response to fish presence and vegetation complexity in western boreal wetlands, with implications for waterbird productivity. Wetlands 26: 1–12.

Hudec, K., and K. Stastny. 1978. Birds in the reedswamp ecosystem. In D. Dykyjová and J. Kvet, eds. *Pond Littoral Ecosystems*, Springer-Verlag, Berlin, pp. 366–375.

Huenneke, L. F., and R. R. Sharitz. 1986. Microsite abundance and distribution of woody seedlings in a South Carolina cypress-tupelo swamp. American Midland Naturalist 115: 328–335.

Huffman, R. T., and R. E. Lonard. 1983. Successional patterns on floating vegetation mats in a southwestern Arkansas bald cypress swamp. Castanea 48: 73–78.

Huston, M. A. 1994. *Biological Diversity: The Coexistence of Species on Changing Landscapes*. Cambridge University Press, Cambridge, England.

Hutchinson, G. E. 1973. Eutrophication: the scientific background of a contemporary practical problem. American Scientist 61: 269–279.

Huttunen, J. T., H. Nykänen, J. Turunen, and P. J. Martikainen. 2003. Methane emissions from natural peatlands in the northern boreal zone in Finland, Fennoscandia. Atmospheric Environment 37: 147–151.

Isselin-Nondedeu, F., L. Rochefort, and M. Poulin. 2007. Long-term vegetation monitoring to assess the restoration success of a vacuum-mined peatland (Québec, Canada). In *International Conference Peat and Peatlands 2007*, pp. 153–166.

Jauhiainen, J., H. Takahashi, J. E. P. Heikkinen, P. J. Martikainen, and H. Vasander. 2005. Carbon fluxes from a tropical peat swamp forest floor. Global Change Biology 11: 1788–1797.

Johnson, F. L., and D. T. Bell. 1976. Plant biomass and net primary production along a flood-frequency gradient in a streamside forest. Castanea 41:break 156–165.

Johnston, C. A., S. D. Bridgham, and J. P. Schubauer-Berigan. 2001. Nutrient dynamics in relation to geomorphology of riverine wetlands. Soil Science Society of America Journal 65: 557–577.

Johnston, C. A., B. Bedford, M. Bourdaghs, T. Brown, C. Frieswyk, M. Tulbere, L. Vaccaro, and J. B. Zedler. 2007. Plant species indicators of physical environment in Great Lakes coastal wetlands. Journal of Great Lakes Research 33: 106–124.

Jones, D. A. 1984. Crabs of the mangal ecosystem. In F. D. Por and I. Dor, eds. *Hydrobiology of the Mangal*. Junk, The Hague, Netherlands, pp. 89–110.

Jørgensen, S. E., and G. Bendoricchio. 2001. *Fundamentals of Ecological Modelling*, 3rd edition, Elsevier, Amsterdam, 530 pp.

Junk, W. J. 1970. Investigations on the ecology and production biology of the 'floating meadows' (*Paspalo echinochloetum*) on the middle Amazon: 1. The floating vegetation and its ecology. Amazonia 2: 449–495.

Kadlec, R. H., and R. L. Knight. 1996. *Treatment Wetlands*. CRC/Lewis Press, Boca Raton, FL, 893 pp.

Kadlec, R.H., and S. Wallace. 2008. *Treatment Wetlands*, 2nd ed., CRC Press, Boca Raton, FL, 1048 pp.

Kang, H., C. Freeman, D. Lee, and W. J. Mitsch. 1998. Enzyme activities in constructed wetlands: Implication for water quality amelioration. Hydrobiologia 368: 231–235.

Keddy, P. A., and A. A. Reznicek. 1986. Great Lakes vegetation dynamics: the role of fluctuating water levels and buried seeds. J. Great Lakes Research 12: 25–36.

Kelley C. A., C. S. Martens, and W. Ussler III. 1995. Methane dynamics across a tidally influenced riverbank margin. Limnology & Oceanography 40: 1112–1129.

Kitchens, W. M., Jr., J. M. Dean, L. H. Stevenson, and J. M. Cooper. 1975. The Santee Swamp as a nutrient sink. In F. G. Howell, J. B. Gentry, and M. H. Smith, eds. Mineral Cycling in Southeastern Ecosystems, ERDA Symposium Series 740513. USGPO. Washington, DC, pp. 349–366.

Kivinen, E., and PP. Pakarinen. 1981. Geographical distribution of peat resources and major peatland complex types in the world, Annals Acad Sciencia Fennicae, Series A, Geology Geography 132: 1–28.

Klopatek, J. M. 1974. *Production of Emergent Macrophytes and Their Role in Mineral Cycling Within a Freshwater Marsh*. Master's Thesis, University of Wisconsin, Milwaukee.

Klopatek J. M. 1978. Nutrient dynamics of freshwater riverine marshes and the role of emergent macrophytes. In R. E. Good, D. F. Whigham, and R. L. Simpson, eds. *Freshwater Wetlands: Ecological Processes and Management Potential*. Academic Press, New York, pp. 195–216.

Knight, J. M., P. E. R. Dale, R. J. K. Dunn, G. J. Broadbent, and C. J. Lemckert. 2008. Patterns of tidal flooding within a mangrove forest: Coombabah Lake, Southeast Queensland. Estuarine, Coastal and Shelf Science 76: 580–593.

Koch M. S., I. A. Mendelssohn, and K. L. McKee. 1990. Mechanism for the hydrogen sulfide-induced growth limitation in wetland macrophytes. Limnology & Oceanography 35: 399–408.

Koerselman, W., and A. F. M. Meuleman. 1996. The vegetation N:P ratio: A new tool to detect the nature of nutrient limitation. Journal of Applied Ecology 33: 1441–1450.

Koreny, J. S., W. J. Mitsch, E. S. Bair, and X. Wu. 1999. Regional and local hydrology of a constructed riparian wetland system. Wetlands 19: 182–193.

Kovacic, D. A., M. B. David, L. E. Gentry, K. M. Starks, and R. A. Cooke. 2000. Effectiveness of constructed wetlands in reducing nitrogen and phosphorus export from agricultural tile drainage. Journal of Environmental Quality 29: 1262–1274.

Krauss, K. W., T. W. Doyle, R. R. Twilley, T. J. Smith, III, K. R. T. Whelan, and J. K. Sullivan. 2005. Woody debris in the mangrove forests of south Florida. Biotropica 37: 9–15.

Kreeger, D. A., and R. I. E. Newell. 2000. Trophic complexity between producers and invertebrate consumers in salt marshes. In M. P. Weinstein and D. A. Kreeger, eds. *Concepts and Controversies in Tidal Marsh Ecology*. Kluwer Academic Publishers, Netherlands, pp. 187–220.

Kremenetski, K. V., A. A. Velichko, O. K. Borisova, G. M. MacDonald, L. C. Smith, K. E. Frey, and L. A. Orlova. 2003. Peatlands of Western Siberian lowlands: current knowledge on zonation, carbon content and Late Quarternary history. Quarternary Science Review 22: 703–723.

Kristensen, E., and D. M. Alongi. 2006. Control by fiddler crabs (*Uca vocans*) and plant roots (*Avicennia marina*) on carbon, iron, and sulfur biogeochemistry in mangrove sediment. Limnology & Oceanography 51: 1557–1571.

Kulesza, A. E., and J. R. Holomuzki. 2006. Amphipod performance responses to decaying leaf litter of *Phragmites australis* and *Typha angustifolia* from a Lake Erie coastal marsh. Wetlands 26: 1079–1088.

Kulczyński, S. 1949. Peat Bogs of Polesie, Acad. Poll Sci. Mem., Series B, No. 15, 356 pp.

Kurz, H., and D. Demaree. 1934. Cypress buttresses in relation to water and air. Ecology 15: 36–41.

Kurz. H., and K. A. Wagner. 1953. Factors in cypress dome development. Ecology 34: 17–164.

Kvet, J., and S. Husak. 1978. Primary data on biomass and production estimates in typical stands of fishpond littoral plant communities. In D. Dykyjova and J. Kvet, eds. *Pond Littoral Ecosystems*. Springer-Verlag, Berlin, pp. 211–216.

LaBaugh, J. W. 1989. Chemical characteristics of water in northern prairie wetlands. In A. van der Valk, ed. *Northern Prairie Wetlands*. Iowa State University Press, Ames, IA, pp. 56–990.

LaBaugh, J. W., T. C. Winter, G. A. Swanson, D. O. Rosenberry, R. D. Nelson, and N. H. Euliss. 1996. Changes in atmospheric circulation patters affect midcontinent wetlands sensitive to climate. Limnology & Oceanography 41: 864–870.

Laderman, A. D. 1989. The Ecology of the Atlantic White Cedar Wetlands: A Community Profile, Biol. Report 8S (7.21), U.S. Fish and Wildlife Service, Washington, DC. 114 pp.

Laderman, A. D., ed. 1998. *Coastally Restricted Forests*. Oxford University Press, Oxford, UK, 334 pp.

Lafleur, P. M., N. T. Roulet, J. L. Bubier, S. Frolking, and T. R. Moore. 2003. Interannual variability in the peatland-atmosphere carbon dioxide exchange at an ombrotrophic bog. Global Biogeochemical Cycles 17: 1036.

Larson, A. C., L. E. Gentry, M. B. David, R. A. Cooke, and D. A. Kovacic. 2000. The role of seepage in constructed wetlands receiving tile drainage. Ecological Engineering 15: 91–104.

Leck, M. A., and K. J. Graveline. 1979. The seed bank of a freshwater tidal marsh. American Journal of Botany 66: 1006–1015.

Leck, M. A., and R. L. Simpson. 1987. Seed bank of a freshwater tidal wetland: turnover and relationship to vegetation change. American Journal of Botany 74: 360–370.

Leck, M. A., and R. L. Simpson. 1995. Ten-year seed bank and vegetation dynamics of a tidal freshwater marsh. American Journal of Botany 82: 1547–1557.

Lee, J. K., R. A. Park, and P. W. Mausel. 1991. Application of geoprocessing and simulation modeling to estimate impacts of sea level rise on northeastern coast of Florida. Photogrammetric Engineering and Remote Sensing 58: 1579–1586.

Lee, S. Y. 1995. Mangrove outwelling: A review. Hydrobiologia 295: 203–212.

Lefeuvre, J. C., and R. F. Dame. 1994. Comparative studies of salt marsh processes in the New and Old Worlds: An introduction. In W. J. Mitsch, ed. *Global Wetlands: Old World and New.* Elsevier, Amsterdam, pp. 169–179.

Lefeuvre, J. C., P. Laffaille, E. Feunteun, V. Bouchard, and A. Radureau. 2003. Biodiversity in salt marshes: From patrimonial value to ecosystem functioning. The case study of Mont-Saint-Michel Bay. C.R. Biologies 326: S125-S131.

Lefeuvre J. C., V. Bouchard, E. Feunteun, S. Grare, P. Lafaille, and A. Radureau. 2000. European salt marshes diversity and functioning: The case study of the Mont Saint-Michel Bay, France. Wetlands Ecology and Management, 8: 147–161.

Lieth, H. 1975. Primary production of the major units of the world. In H. Lieth and R. H. Whitaker, eds. *Primary Productivity of the Biosphere.* Springer-Verlag, New York, pp. 203–215.

Light, H. M., M. R. Durst, L. J. Lewis, and D. A. Howell. 2002. Hydrology, Vegetation, and Soils of Riverine and Tidal Floodplain Forests of the Lower Suwanee River, Florida and Potential Impacts of Flow Reductions. U.S. Geological Survey Professional Paper 1656A, 124 pp.

Likens, G. E., F. H. Bormann, R. S. Pierce, J. S. Eaton, and N. M. Johnson. 1977. *Biogeochemistry of a Forested Ecosystem.* Springer-Verlag, New York.

Lindeman, R. L. 1942. The trophic-dynamic aspect of ecology. Ecology 23: 399–418.

Lipschultz, F. 1981. Methane release from a brackish intertidal salt-marsh embayment of Chesapeake Bay, Maryland. Estuaries 4: 143–145.

Little, E. L., Jr. 1971. Atlas of United States Trees, vol. 1, Conifers and important hardwoods, Misc. Pub. No. 1146, U.S. Department of Agriculture—Forest Service, USGPO, Washington, DC.

Lockaby, B. G., and M. R. Walbridge. 1998. Biogeochemistry. In M.G. Messina and W.H. Conner, eds. *Southern Forested Wetlands: Ecology and Management.* Lewis Publisher, Boca Raton, FL, pp. 149–172.

Locky, D. L., S. E. Bayley, and D. H. Vitt. 2005. The vegetational ecology of black spruce swamps, fens, and bogs in southern boreal Manitoba, Canada. Wetlands 25: 564–582.

Lovelock, C. E. 2008. Soil respiration and belowground carbon allocation in mangrove forests. Ecosystems 11: 342–354.

Lugo, A. E. 1980. Mangrove ecosystems: successional or steady state? Biotropica (supplement) 12: 65–72.

Lugo, A. E. 1988. The mangroves of Puerto Rico are in trouble. Acta Cientifica 2: 124.

Lugo, A. E., G. Evink, M. M. Brinson, A. Broce, and J. C. Snedaker. 1975. Diurnal rates of photosynthesis, respiration, and transpiration in mangrove forests in South Florida. In F. B. Golley and E. Medina, eds. *Tropical Ecological Systems—Trends in Terrestrial and Aquatic Research*. Springer-Verlag, New York, pp. 335–350.

Lugo, A. E., and C. Patterson-Zucca. 1977. The impact of low temperature stress on mangrove structure and growth. Tropical Ecology 18: 149–161.

MacIntyre, H. L., and J. J. Cullen. 1995. Fine-scale vertical resolution of chlorophyll and photosynthetic parameters in shallow-water benthos. Marine Ecology Progress Series 122: 227–237.

Malecki-Brown, L. M., J. R. White, and K. R. Reddy. 2007. Soil biogeochemical characteristics influenced by alum application in a municipal wastewater treatment wetland. Journal of Environmental Quality 36: 1904–1913.

Malley, D. F. 1978. Degradation of mangrove leaf litter by the tropical sesarmid crab *Chiromanthes onychophorum*. Marine Biology 49: 377–386.

Malmer, N. 1975. Development of bog mires. In A. D. Hasler, ed. *Coupling of Land and Water Systems*. Springer-Verlag, New York, pp. 85–92.

Martin, J. F., and D. R. Tilley. 2000. Simulating with STELLA. In H. T. Odum and E. C. Odum, eds. *Modeling for All Scales: An Introduction to System Simulation*. Academic Press, San Diego, pp. 133–150.

Martinez, M. L., A. Intralawan, G. Vázquez, O. Péres-Maqueo, P. Sutton, and R. Landgrave. 2007. The coasts of the world: Ecological, economic and social importance. Ecological Economics 63: 254–272.

Matthews, E., and I. Fung. 1987. Methane emissions from natural wetlands: global distribution, area, and environmental characteristics of sources. Global Biogeochemical Cycles 1: 61–86.

McIntosh, R. P. 1985. *The Background of Ecology: Concept and Theory*. Cambridge University Press, Cambridge, 383 pp.

McJannet, C. L., P. A. Keddy, and F. R. Pick. 1995. Nitrogen and phosphorus tissue concentrations in 41 wetland plants: A comparison across habitats and functional groups. Functional Ecology 9: 231–238.

McKee, K. L., and J. E. Rooth. 2008. Where temperate meets tropical: Multi-factorial effects of elevated CO_2, nitrogen enrichment, and competition on a mangrove-salt marsh community. Global Change Biology 14: 971–984.

McLaughlin, D. B., and H. J. Harris. 1990. Aquatic insect emergence in two Great Lakes marshes. Wetlands Ecology and Management 1: 111–121.

McNaughton, S. J., 1966. Ecotype function in the *Typha* community-type. Ecological Monographs 36: 297–325.

McNaughton, S. J., and L. L. Wolf. 1973. *General Ecology*. Holt, Renehart and Winston, Inc., New York.

Megonigal, J. P., and F. P. Day, Jr. 1988. Organic matter dynamics in four seasonally flooded forest communities of the Dismal Swamp. American Journal of Botany 75: 1334–1343.

Megonigal, J. P., W. H. Conner, S. Kroeger, and R. R. Sharitz. 1997. Aboveground production in Southeastern floodplain forests: A test of the subsidy-stress hypothesis. Ecology 78: 370–384.

Mendelssohn, I. A., and J. T. Morris. 2000. Eco-physiological controls on the productivity of *Spartina alterniflora* Loisel. In M. P. Weinstein and D. A. Kreeger, eds. *Concepts and Controversies in Tidal Marsh Ecology*. Kluwer Academic Publishers, The Netherlands, pp. 59–80.

Metzker, K. D., and W. J. Mitsch. 1997. Modelling self-design of the aquatic community in a newly created freshwater wetland. Ecological Modelling 100: 61–86.

Micheli, F., F. Gherardi, and M. Vannini. 1991. Feeding and burrowing ecology of two East African mangrove crabs. Marine Biology 111: 247–254.

Middleton, B. A. 1999. *Wetland Restoration: Flood Pulsing and Disturbance Dynamics*. John Wiley & Sons, Inc., New York, 388 pp.

Mitchell, D. S., and B. Gopal. 1991. Invasion of Tropical Freshwaters by Alien Aquatic Plants. In P. S. Ramakrishnan, ed., *Ecology of Biological Invasion in the Tropics*. pp. 139–154.

Mitsch, W. J. 1979. Interactions between a riparian swamp and a river in southern Illinois. In R. R. Johnson and J. F. McComlick, tech. coords. *Strategies for the Protection and Management of Floodplain Wetlands and Other Riparian Ecosystems*, Proceedings of a Symposium, Calaway Gardens, U.S. Forest Service General Technical Report WO-12, Washington, DC, pp. 63–72.

Mitsch, W. J. 1988. Productivity-hydrology-nutrient models of forested wetlands. In W. J. Mitsch, M. Straskraba and S. E. Jørgensen, eds. *Wetland Modelling*. Elsevier, Amsterdam, pp. 115–132.

Mitsch, W. J., and K. C. Ewel. 1979. Comparative biomass and growth of cypress in Florida wetlands. American Midland Naturalist 101: 417–426.

Mitsch, W. J., C. L. Dorge, and J. R. Wiemhoff. 1979. Ecosystem dynamics and a phosphorus budget of an alluvial cypress swamp in southern Illinois. Ecology 60: 1116–1124.

Mitsch, W. J., J. W. Day, Jr., J. R. Taylor, and C. Madden. 1982. Models of North American freshwater wetlands. International Journal of Ecological and Environmental Science 8: 109–140.

Mitsch, W. J., and W. G. Rust. 1984. Tree growth responses to flooding in a bottomland forest in northeastern Illinois. Forest Science 30: 499–510.

Mitsch, W. J., and J. G. Gosselink. 1986. *Wetlands*. Van Nostrand Reinhold, New York.

Mitsch, W. J., and B. C. Reeder. 1991. Modelling nutrient retention of a freshwater coastal wetland: estimating the roles of primary productivity, sedimentation, resuspension and hydrology. Ecological Modelling 54: 151–187.

Mitsch, W. J., J. R. Taylor, and K. B. Benson. 1991. Estimating primary productivity of forested wetland communities in different hydrologic landscapes. Landscape Ecology 5: 75–92.

Mitsch, W. J., and J. G. Gosselink. 1993. *Wetlands*, 2nd ed., John Wiley & Sons, Inc., New York, 722 pp.

Mitsch, W. J., J. K. Cronk, X. Wu, R. W. Nairn, and D. L. Hey. 1995. Phosphorus retention in constructed freshwater riparian marshes. Ecological Applications 5: 830–845.

Mitsch, W. J., and K. M. Wise. 1998. Water quality, fate of metals, and predictive model validation of a constructed wetland treating acid mine drainage. Water Research 32: 1888–1900.

Mitsch, W. J., X. Wu, R. W. Nairn, P. E. Weihe, N. Wang, R. Deal, and C. E. Boucher. 1998. Creating and restoring wetlands: A whole-ecosystem experiment in self-design. BioScience 48: 1019–1030.

Mitsch, W. J., and J. G. Gosselink. 2000. *Wetlands*, 3rd ed., John Wiley & Sons, Inc., New York, 920 pp.

Mitsch, W. J., A. J. Horne, and R. W. Nairn. 2000. Nitrogen and phosphorus retention in wetlands: Ecological approaches to solving excess nutrient problems. Ecological Engineering 14: 1–7.

Mitsch, W. J., and J. W. Day, Jr. 2004. Thinking big with whole ecosystem studies and ecosystem restoration—A legacy of H.T. Odum. Ecological Modelling 178: 133–155.

Mitsch, W. J., and S. E. Jørgensen. 2004. *Ecological Engineering and Ecosystem Restoration*. John Wiley & Sons, Inc., New York, 411 pp.

Mitsch, W. J., N. Wang, L. Zhang, R. Deal, X. Wu, and A. Zuwerink. 2005a. Using ecological indicators in a whole-ecosystem wetland experiment. In S.E. Jørgensen, F-L. Xu, and R. Costanza, eds. *Handbook of Ecological Indicators for Assessment of Ecosystem Health*. CRC Press, Boca Raton, FL, pp. 211–235.

Mitsch, W. J., L. Zhang, C. J. Anderson, A. Altor, and M. Hernandez. 2005b. Creating riverine wetlands: Ecological succession, nutrient retention, and pulsing effects. Ecological Engineering 25: 510–527.

Mitsch, W. J., and J. G. Gosselink. 2007. *Wetlands*, 4th ed., John Wiley & Sons, Inc., Hoboken, NJ, 582 pp.

Mitsch, W. J., R. H. Mitsch, and R. E. Turner. 1994. Wetlands of the Old and New Worlds—ecology and management. In: W. J. Mitsch, ed. Global Wetlands: Old World and New, Elsevier, Amsterdam, pp. 3–56.

Montague, C. L., and R. G. Wiegert. 1990. Salt marshes. In R. L. Myers and J. J. Ewel, eds. *Ecosystems of Florida*. University of Central Florida Press, Orlando, FL, pp. 481–516.

Moore, D. R. J., and P. A. Keddy. 1989. The relationship between species richness and standing crop in wetlands: The importance of scale. Vegetatio 79: 99–106.

Moore, D. R. J., P. A. Keddy, C. L. Gaudet, and I. C. Wisheu. 1989. Conservation of wetlands: do infertile wetlands deserve a higher priority? Biological Conservation 47: 203–217.

Moore, P. D., and D. J. Bellamy. 1974. *Peatlands*. Springer-Verlag, New York, 221 pp.

Moore, T. R. 1989. Growth and net production of *Sphagnum* at five fen sites, subarctic eastern Canada. Canadian Journal of Botany 67: 1203–1207.

Morris, J. T., and W. B. Bowden. 1986. A mechanistic, numerical model of sedimentation, mineralization, and decomposition for marsh sediments. Journal of Soil Science Society of America 50: 96–105.

Morris, J. T., B. Kjerfve, and J. M., Dean. 1990. Dependence of estuarine productivity on anomalies in mean sea level. Limnology & Oceanography 35: 926–930.

Morris, J. T., P. V. Sundareshwar, C. T. Nietch, B. Kjerfve, and D. R. Cahoon. 2002. Response of coastal wetlands to rising sea level. Ecology 83: 2869–2877.

Motzkin, G., W. A. Patterson, and N. E. R. Drake. 1993. Fire history and vegetation dynamics of a *Chamaecyparis thyoides* wetland on Cape Cod, Massachusetts. Journal of Ecology 81: 391–402.

Mulholland, P. J. 1979. Organic Carbon in a Swamp-Stream Ecosystem and Export by Streams in Eastern North Carolina, Ph.D. Dissertation, University of North Carolina, Chapel Hill.

Murdock, N. A. 1994. Rare and endangered plants and animals of southern Appalachian wetlands. Water, Air, and Soil Pollution 77: 385–405.

Murkin, H. R., A. G. van der Valk, and C. B. Davis. 1989. Decomposition of four dominant macrophytes in the Delta Marsh, Manitoba. Wildlife Society Bulletin 17: 215–221.

Nahlik, A. M., and W. J. Mitsch. 2008. The effect of river pulsing on sedimentation and nutrients in created riparian wetlands. Journal of Environmental Quality 37: 1634–1643.

Naiman, R. J., T. Manning, and C. A. Johnston. 1991. Beaver population fluctuations and troposheric methane emissions in boreal wetlands. Biogeochemistry 12: 1–15.

Nairn, R. W., and W. J. Mitsch. 2000. Phosphorus removal in created wetland ponds receiving river overflow. Ecological Engineering 14: 107–126.

National Research Council (NRC). 1995. *Wetlands: Characteristics and Boundaries*. National Academy Press, Washington, DC, 306 pp.

Naumann, E. 1919. Nagra sypunkte angaende planktons ökilogi. Med. sarskild hänsyn till fytoplankton. Svensk Botanisk Tidkrift 13: 129–158.

Neckles, H. A., H. R. Murkin, and J. A. Cooper. 1990. Influences of seasonal flooding on macroinvertebrate abundance in wetland habitats. Freshwater Biology 23: 311–322.

Neill, C. 1990a. Effects of nutrients and water levels on emergent macrophytes biomass in a prairie marsh, Canadian Journal of Botany 68: 1007–1014.

Neil, C. 1990b. Effects of nutrients and water levels on species composition in prairie whitetop (*Scolochloa festucacea*) marshes. Canadian Journal of Botany 68: 1015–1020.

Neill, C. 1990c. Nutrient limitation of hardstem bulrush (*Scirpus acutus* Muhl.) in a Manitoba interlake region marsh. Wetlands 10: 69–75.

Nelson, J. W., J. A. Kadlec, and H. R. Murkin. 1990a. Seasonal comparisons of weight loss of two types of *Typha glauca* Godr. leaf litter. Aquatic Biology 37: 299–314.

Nelson, J. W., J. A. Kadlec, and H. R. Murkin. 1990b. Response by macroinvertebrates to cattail litter quality and timing of litter submergence in a northern prairie marsh. Wetlands 10: 47–60.

Nessel, J. K. 1978. Distribution and Dynamics of Organic Matter and Phosphorus in a Sewage Enriched Cypress Strand, M.S. Thesis, University of Florida, Gainesville, FL, 159 pp.

Neubauer, S. C., K. Givler, S. Valentine, and J. P. Megonigal. 2005. Seasonal patterns and plant-mediated controls of subsurface wetland biogeochemistry. Ecology 86: 3334–3344.

Newell, S. Y., and D. Porter. 2000. Microbial secondary production from saltmarsh grass shoots, and its known and potential fates. In M. P. Weinstein and D. A. Kreeger, eds. *Concepts and Controversies in Tidal Marsh Ecology*. Kluwer Academic Publishers, Dordrecht, Netherlands, pp. 159–185.

Nichols, D. S. 1983. Capacity of natural wetlands to remove nutrients from wastewater. Journal of Water Pollution Control Federation 55: 495–505.

Nicholson, B. J., L. D. Gignac, and S. E. Bayley. 1996. Peatland distribution along a north-south transect in the Mackenzie River Basin in relation to climatic and environmental gradients. Vegetatio 126: 119–133.

Nixon, S. W. 1980. Between coastal marshes and coastal waters—a review of twenty years of speculation and research on the role of salt marshes in estuarine productivity and water chemistry. In PP. Hamilton and K. B. MacDonald, eds. *Estuarine and Wetland Processes*. Plenum, New York, pp. 437–525.

Noe, G. P., and C. R. Hupp. 2005. Carbon, nitrogen, and phosphorus accumulation in floodplains of Atlantic Coastal Plain rivers, USA. Ecological Applications 15: 1178–1190.

Noormets, A., J. Chen, S. D. Bridgham, J.F. Weltzin, J. Pastor, B. Dewey, and J. LeMoine. 2004. The effects of infrared loading and water table on soil energy fluxes in northern peatlands. Ecosystems 7: 573–582.

Novitzki, R. P. 1979. Hydrologic characteristics of Wisconsin's wetlands and their influence on floods, stream flow, and sediment. In P. E. Greeson, J. R. Clark, J. E. Clark eds. *Wetland Functions and Values: The State of Our Understanding*. American Water Resource Association, Minneapolis, MN, pp. 377–388.

Novitzki, R. P. 1982. Hydrology of Wisconsin Wetlands. University of Wisconsin Extension Geological Natural History Survey Circular 40, University of Wisconsin, Madison, 27 pp.

Odum, E. P. 1968. A research challenge: Evaluating the productivity of coastal and estuarine water. Proceedings Second Sea Grant Conference, University of Rhode Island, pp. 63–64.

Odum, E. P. 1969. The strategy of ecosystem development. Science 164: 262–270.

Odum, E. P. 1980. The status of three ecosystem-level hypotheses regarding salt marsh estuaries: tidal subsidy, outwelling, and detritus-based food chains. In V. S. Kennedy, ed. *Estuarine Perspectives*. Academic Press, New York, pp. 485–495.

Odum, E. P. 1984. The mesocosms. BioScience 34: 558–562.

Odum, E. P. 2000. Tidal marshes as outwelling/pulsing systems. In M. P. Weinstein and D. A. Kreeger, eds. *International Symposium: Concepts and Controversies in Tidal Marsh Ecology*. Kluwer Academic Publishers, Dordrecht, Netherlands, pp. 3–7.

Odum. E. P., and A. A. de la Cruz. 1967. Particulate organic detritus in a Georgia salt marsh-estuarine ecosystem. In G. H. Lauer, ed. *Estuaries*. American Association for the Advancement of Science, Washington, DC, pp. 383–388.

Odum, E. P., J. T. Finn, and E. H. Franz. 1979. Perturbation theory and the subsidy-stress gradient. BioScience 29: 349–352.

Odum, H. T. 1982. Role of wetland ecosystems in the landscape of Florida. In D. O. Logofet and N. K Luckyanov, eds. *Ecosystem Dynamics in Freshwater Wetlands and Shallow Water Bodies*, Vol. II. SCOPE and UNEP Workshop, Center of International Projects, Moscow, USSR, pp. 33–72.

Odum, H. T. 1983, *Systems Ecology: An Introduction*. John Wiley & Sons, New York, 644 pp.

Odum, H. T. 1984. Summary: cypress swamps and their regional role. In K. C. Ewel and H. T. Odum, eds. *Cypress Swamps*. University Presses of Florida, Gainesville, pp. 416–443.

Odum, H. T. 1985. Self Organization of Ecosystems in Marine Ponds Receiving Treated Sewage. UNC Sea Grant SG-85-04, 250 pp.

Odum, H. T. 1989. Experimental study of self organization in estuarine ponds. In W. J. Mitsch and S. E. Jørgensen, eds. *Ecological Engineering*. John Wiley & Sons, New York, pp. 291–340.

Odum. H. T., and R. F. Pigeon, eds. 1970. *A Tropical Rain Forest*, 3 volumes. Office of Information Services, U.S. Atomic Energy Commission, NTIS, Springfield, VA.

Odum, H. T., K. C. Ewel, W. J. Mitsch, and J. W. Ordway. 1977. Recycling treated sewage through cypress wetlands in Florida. In F. M. D'Itri, ed. *Wastewater Renovation and Reuse*. Marcel Dekker, Inc., New York, pp. 35–67.

Odum, H. T., and E. C. Odum. 2000. *Modeling for All Scales: An Introduction to System Simulation*. Academic Press, San Diego, 458 pp.

Odum, W. E. 1970. Pathways of Energy Flow in a South Florida Estuary, Ph.D. Dissertation, University of Miami, Coral Gables, Florida, 1–62 pp.

Odum, W. E., and E. J. Heald. 1972. Trophic analyses of an estuarine mangrove community. Bulletin of Marine Science 22: 671–738.

Odum, W. E., C. C. McIvor, and T. J. Smith. 1982. The Ecology of the Mangroves of South Florida: a Community Profile. U.S. Fish & Wildlife Service, Office of Biological Services, Technical Report FWS/OBS 81-24, Washington, DC.

Odum, W. E, T. J. Smith III, J. K Hoover, and C. C. McIvor. 1984. The Ecology of Freshwater Marshes of the United States East Coast: A Community Profile. U.S. Fish and Wildlife Service, FWS/OBS-87/17, Washington, DC, 177 pp.

Odum, W. E., E. P. Odum, and H. T. Odum, 1995. Nature's pulsing paradigm. Estuaries 18: 547–555.

Oliver, J. D., and T. Legovic. 1988. Okefenokee marshland before, during, and after nutrient enrichment by a bird rookery. Ecological Modelling 43: 195–223.

Osborne, K., and T. J. Smith. 1990. Differential predation on mangrove propagules in open and closed canopy forest habitats. Vegetatio 89: 1–6.

Ostendorp, W., C. Iseli, M. Krauss, P. Krumscheid-Plankert, J. L. Moret, M. Rollier, and F. Schanz. 1995. Lake shore deterioration, reed management and bank restoration in some Central European lakes. Ecological Engineering 5: 51–75.

Overbeck, F., and H. Happach. 1957. Über das Wachstum und den Wasserhaushalt einiger Hochmoor Sphagnum, Flora. Jena 144: 335–402.

Ozalp, M., W. H. Conner, and B. G. Lockaby. 2007. Above-ground productivity and litter decomposition in a tidal freshwater forested wetland on Bull Island, SC, USA. Forest Ecology and Management 245: 31–43.

Özesmi, U., and W. J. Mitsch. 1997. A spatial model for the marsh-breeding red-winged blackbird (*Agelaius phoeniceus* L.) in coastal Lake Erie wetlands. Ecological Modelling 101: 139–152.

Pakarinen, P. 1978. Production and nutrient ecology of three *Sphagnum* species in southern Finnish raised bogs, Annales Botanici Fennici 15: 15–26.

Pakarinen, P., and E. Gorham. 1983. Mineral Element Composition of *Sphagnum* fuscum peats collected from Minnesota, Manitoba, and Ontario. In Proceedings of the International Symposium on Peat Utilization, Bemidji State Univ. Center for Environmental Studies, Bemidji, MN, pp. 417–429.

Pallis, M. 1915. The structural history of Plav: the floating fen of the delta of the Danube. J. Linn. Soc. Bot. 43: 233–290.

Park, R. A., J. K. Lee, PP. W. Mausel, and R. C. Howe. 1991. Using remote sensing for modelling the impacts of sea level rise. World Resource Review 3: 184–205.

Patten, B. C., and S. E. Jørgensen. 1995. *Complex Ecology: The Part-Whole Relation in Ecosystems*. Prentice Hall, Englewood Cliffs, NJ, 705 pp.

Patterson, S. G. 1986. Mangrove Community Boundary Interpretation and Detection of Areal Changes in Marco Island, Florida: Application of Digital Image Processing and Remote Sensing Techniques. Biological Services Report 86 (10), U.S. Fish and Wildlife Service, Washington, DC.

Pearlstein, L., H. McKellar, and W. Kitchens. 1985. Modelling the impacts of a river diversion on bottomland forest communities in the Santee River floodplain, South Carolina, *Ecological Modelling* 29: 283–302.

Pedersen, A. 1975. Growth measurements of five *Sphagnum* species in South Norway. *Norwegian Journal of Botany* 22: 277–284.

Pelikan, J. 1978. Mammals in the reedswamp ecosystem. In D. Dykyjova and J. Kvet, eds. *Pond Littoral Ecosystems*. Springer-Verlag, Berlin, pp. 357–365.

Penfound, W. T. 1952. Southern swamps and marshes, Botanical Review 18: 413–446.

Penfound, W. T., and T. T. Earle. 1948. The biology of the water hyacinth, Ecological Monographs 18: 447–472.

Perry, J. E., and C. H. Hershner. 1999. Temporal changes in the vegetation pattern in a tidal freshwater marsh. Wetlands 19: 90–99.

Petillon, J., F. Ysnel, A. Canard, and J. C. Lefeuvre. 2005. Impact of an invasive plant (*Ellymus athericus*) on the conservation value of tidal salt marshes in western France and implications for management: Responses of spider populations. Biological Conservation 126: 103–117.

Pfeiffer, W. J., and R. G. Wiegert. 1981. Grazers on *Spartina* and their predators. In L. R. Pomeroy and R. G. Wiegert, eds. *The Ecology of a Salt Marsh*. Springer-Verlag, New York, pp. 87–112.

Philipp, K. R., and R. T. Field. 2005. *Phragmites australis* expansion in Delaware Bay salt marshes. Ecological Engineering 25: 275–291.

Phipps, R. G., and W. G. Crumpton. 1994. Factors affecting nitrogen loss in experimental wetlands with different hydrologic loads. Ecological Engineering 3: 399–408.

Phipps, R. L. 1979. Simulation of wetlands forest vegetation dynamics. Ecological Modelling 7: 257–288.

Phipps, R. L., and L. H. Applegate. 1983. Simulation of management alternatives in wetland forests. In S. E. Jørgensen and W. J. Mitsch, eds. *Application of Ecological Modelling in Environmental Management*, Part B. Elsevier, Amsterdam, pp. 311–339.

Pinckney, J., and R. G. Zingmark. 1993. Modeling the annual production of intertidal benthic microalgae in estuarine ecosystems. Journal of Phycology 29: 396–407.

Pjavchenko, N. J. 1982. Bog ecosystems and their importance in nature. In D. O. Logofet and N. K. Luckyanov, eds. Proceedings of International Workshop on Ecosystems Dynamics in Wetlands and Shallow Water Bodies, Vol. 1, SCOPE and UNEP Workshop. Center for International Projects, Moscow, USSR, pp. 7–21.

Pollock, M. M., R. J. Naiman, and T. A. Hanley. 1998. Plant species richness in riparian wetlands—a test of biodiversity theory. Ecology 79: 94–105.

Pomeroy, L. R., L. R. Shenton, R. D. Jones, and R. J. Reimold. 1972. Nutrient flux in estuaries. In G. E. Likens, ed. *Nutrients and Eutrophication*. American

Society of Limnology and Oceanography Special Symposium. Allen Press, Lawrence, Kansas, pp. 274–291.

Pomeroy, L. R. 1959. Algae productivity in salt marshes of Georgia. Limnology and Oceanography 4: 386–397.

Pomeroy, L. R., W. M. Darley, E. L. Dunn, J. L Gallagher, E. B. Haines, and D. M. Witney. 1981. Primary production. In: L. R Pomeroy and R. G. Wiegert, eds. Ecology of a Salt Marsh. Springer-Verlag, New York, pp. 39–67.

Porej, D. 2004. Faunal aspects of wetland creation and restoration. Ph.D. dissertation, Evolution, Ecology, and Organismal Biology Department, The Ohio State University, Columbus.

Porej, D., M. Micacchion, and T. E. Hetherington. 2005. Core terrestrial habitat for conservation of local populations of salamanders and wood frogs in agricultural landscapes. Biological Conservation 120: 399–409.

Post, R. A. 1996. Functional Profile of Black Spruce Wetlands in Alaska. Report EPA 910/R-96-006, U.S. Environmental Protection Agency, Seattle, WA, 170 pp.

Powell, S. W., and F. P. Day. 1991. Root production in four communities in the Great Dismal Swamp. American Journal of Botany 78: 288–297.

Price, J. S., L. Rochefort, and F. Quinty. 1998. Energy and moisture considerations on cutover peatlands: Surface microtopography, mulch cover, and *Sphagnum* regeneration. Ecological Engineering 10: 293–312.

Rabinowitz, D. 1978. Dispersal properties of mangrove propagules, Biotropica 10: 47–57.

Reader, R. J., and J. M. Stewart. 1972. The relationship between net primary production and accumulation for a peatland in southeastern Manitoba. Ecology 53: 1024–1037.

Reimold, R. J. 1974. Mathematical modeling—*Spartina*. In R. J. Rermold and W. M. Queen, eds. *Ecology of Halophytes*, Academic Press, New York, pp. 393–406.

Reiners, W. A. 1972. Structure and energetics of three Minnesota forests. Ecological Monographs 42: 71–94.

Rheinhardt, R. 1992. A multivariate analysis of vegetation patterns in tidal freshwater swamps of lower Chesapeake Bay, USA. Bulletin of the Torrey Botanical Club 119: 192–207.

Richardson, C. J., W. A. Wentz, J. P. M. Chamie, J. A. Kadlec, and D. L. Tilton. 1976. Plant growth, nutrient accumulation and decomposition in a central Michigan peatland used for effluent treatment. In D. L. Tilton, R. H. Kadlec, and C. J. Richardson, eds. *Freshwater Wetlands and Sewage Effluent Disposal*. University of Michigan, Ann Arbor, MI, pp. 77–117.

Richardson, C. J. 2003. Pocosins: hydrologically isolated or integrated wetlands on the landscape? Wetlands 23: 563–576.

Richter, K. O., and A. L. Azous. 1995. Amphibian occurrence and wetland characteristics in the Puget Sound Basin. Wetlands 15: 305–312.

Rivera-Monroy, V. H., J. W. Day, R. R. Twilley, R. Vera-Herrera, and C. Coronado-Molina. 1995. Flux of nitrogen and sediment in a fringe mangrove forest in Terminos Lagoon, Mexico. Estuarine, Coastal, and Shelf Science 40: 139–160.

Rivers, J. S., D. I. Siegel, L. S. Chasar, J. PP. Chanton, PP. H. Glaser, N. T. Roulet, and J. M. McKenzie. 1998. A stochastic appraisal of the annual carbon budget of a large circumboreal peatland, Rapid River Watershed, northern Minnesota. Global Biogeochemical Cycles 12: 715–727.

Robertson, A. I. 1986. Leaf-burying crabs: Their influence on energy flow and export from mixed mangrove forests (*Rhizophera* spp.) in northeastern Australia. Journal of Experimental Marine Biology and Ecology 102: 237–248.

Robertson, A. I., and P. A. Daniel. 1989. The influence of crabs on litter processing in high intertidal mangrove forests in tropical Australia. Oceologia 78: 191–198.

Robertson, A. I., P.Y. Bacon, and G. Heagney. 2001. The response of floodplain primary production to flood frequency and timing. Journal of Applied Ecology 38: 126–136.

Rochefort, L., D. H. Vitt, and S. E. Bayley. 1990. Growth, production, and decomposition dynamics of *Sphagnum* under natural and experimentally acidified conditions. Ecology 71: 1986–2000.

Rochefort, L., G. Quinty, S. Campeau, K. Johnson, and T. Malterer. 2003. North American approach to the restoration of *Sphagnum* dominated peatlands. Wetlands Ecology and Management 11: 3–20.

Romero, L. M., T. J. Smith III, and J. W. Fourqurean. 2005. Changes in mass and nutrient content of wood during decomposition in a south Florida mangrove forest. Journal of Ecology 93: 618–631.

Rose, C., and W. G. Crumpton. 2006. Spatial patterns in dissolved oxygen and methane concentrations in prairie pothole wetlands in Iowa, USA. Wetlands 26: 1020–1025.

Rosswall, T., and O. W. Heal. 1975. Structure and function of tundra ecosystems. Ecological Bulletin (Sweden) 20: 265–294.

Rymal, D. E., and G. W. Folkerts. 1982. Insects associated with pitcher plants (*Sarracenia*, salraceniaceae), and their relationship to pitcher plans conservation: A review. Journal of Alabama Academy of Sciences 53: 131–151.

Saltonstall, K., P.M. Peterson, and R.J. Soreng. 2004. Recognition of *Phragmites australis* subsp. *americanus* (Poaceae: Arundinoideae) in North America: Evidence from morphological and genetic analyses. Brit. Org/SIDA 21: 683–692.

Sanders, M. D., and M. J. Winterbourn. 1993. Effect of *Sphagnum* harvesting on invertebrate species diversity and community size. New Zealand Department of Conservation Science and Research Series No. 57, Wellington.

Sanville, W., and W. J. Mitsch, eds. 1994. Creating Freshwater Marshes in a Riparian Landscape: Research at the Des Plaines River Wetland Demonstration Project. Special Issue of Ecological Engineering 3: 315–521.

Sasikala, S., N. Tanaka, and K. B. S. N. Jinadasa. 2008. Effect of water level fluctuations on nitrogen removal and plant growth performance in vertical subsurface-flow wetland mesocosms. Journal of Freshwater Ecology 23: 101–112.

Sasser, C. E., G. W. Peterson, D. A. Fuller, R. K. Abernethy, and J. G. Gosselink. 1982. Environmental Monitoring Program, Louisiana Offshore Oil Port Pipeline. 1981. Annual Report, Coastal Ecology Laboratory, Center for Wetland Resources, Louisiana State University, Baton Rouge, 299 pp.

Sasser, C. E., and J. G. Gosselink. 1984. Vegetation and primary production in a floating wetland marsh in Louisiana. Aquatic Botany 20: 245–255.

Sasser, C. E., J. G. Gosselink, and G. P. Shaffer. 1991. Distribution of nitrogen and phosphorus in a Louisiana freshwater floating marsh. Aquatic Botany 41: 317–331.

Sasser, C. E., J. M. Visser, D. E. Evers, and J. G. Gosselink. 1995. The role of environmental variables on interannual variation in species composition and biomass in a subtropical minerotrophic floating marsh. Canadian Journal of Botany 73: 413–424.

Sasser, C. E., J. G. Gosselink, E. M. Swenson, C. M. Swarzenski, and N. C. Leibowitz. 1996. Vegetation, substrate and hydrology in floating marshes in the Mississippi river delta plain wetlands, USA. Vegetatio 122: 129–142.

Schaeffer-Novelli, Y., G. Cintron-Molero, R. R. Adaime, and T. M. de Camargo. 1990. Variability of mangrove ecosystems along the Brazilian coast. Estuaries 13: 204–218.

Schilling, E. B., and B. G. Lockaby. 2006. Relationships between productivity and nutrient circulation within two contrasting southeastern U.S. floodplain forests. Wetlands 26: 181–192.

Schindler, D. E., S. R. Carpenter, J. J. Cole, J. F. Kitchell, and M. L. Pace. 1997. Influence of food web structure on carbon exchange between lakes and the atmosphere. Science 227: 248–251.

Schindler, D. W. 1977. Evolution of phosphorus limitation in lakes. Science 195: 260–262.

Schindler, D. W. 1998. Replication versus realism: The need for ecosystem scale experiments. Ecosystems 1: 323–334.

Schipper, L. A., B. R. Clarkson, M. Vojvodic-Vukovic, and R. Webster. 2002. Restoring cut-over restiad peat bogs: A factorial experiment of nutrients, seed, and cultivation. Ecological Engineering 19: 29–40.

Schlesinger, W. H. 1978. Community structure, dyes, and nutrient ecology in the Okefenokee Cypress Swamp-Forest. Ecological Monographs 48: 43–65.

Schneider, R. L., and R. R. Sharitz. 1986. Seed bank dynamics in a southeastern riverine swamp. American Journal of Botany 73: 1022–1030.

Schneider, R. L. and R. R. Sharitz. 1988. Hydrochory and regeneration in a bald cypress-water tupelo swamp forest. Ecology 69: 1055–1063.

Selbo, S. M., and A. A. Snow. 2004. The potential for hybridization between *Typha angustifolia* and *T. latafolia* in a constructed wetland. Aquatic Botany 78: 361–369.

Sharitz, R. R., and J. W. Gibbons, eds. 1982. The Ecology of Southeastern Shrub Bogs (Pocosins) and Carolina Bays: A Community Profile, U.S. Fish Wildlife Service, Division of Biological Services, Washington, DC, FWS/OBS-82/04.

Shaver, G. R., and J. M. Melillo. 1984. Nutrient budgets of marsh plants: efficiency concepts and relation to availability. Ecology 65: 1491–1510.

Shaw, S. P., and C. G. Fredine. 1956. Wetlands of the United States: Their Extent, and Their Value for Waterfowl and Other Wildlife. U.S. Department of Interior, Fish and Wildlife Service, Circular 39, Washington, DC, 67 pp.

Sheffield, R. M., T. W. Birch, W. H. McWilliams, and J. B. Tansey. 1998. *Chamaecyparis thyoides* (Atlantic White Cedar) in the United States. In Laderman, A.D., ed. *Coastally Restricted Forests*. Oxford University Press, New York, pp. 111–123.

Sherman, R. E., T. J. Fahey, and P. Martinez. 2003. Spatial patterns of biomass and aboveground net primary productivity in a mangrove ecosystem in the Dominican Republic. Ecosystems 6: 384–398.

Siegel, D. I., A. S. Reeve, P. H. Glaser, and E. A. Romanowicz. 1995. Climate-driven flushing of pore water in peatlands. Nature: 531–533.

Silliman, B. R., H. van de Koppel, M. D. Bertness, L. E. Stanton, and I. A. Mendelssohn. 2005. Drought, snails, and large-scale die-off of southern U.S. salt marshes. Science 310: 1803–1806.

Silvola, J., and I. Hanski. 1979. Carbon accumulation in a raised bog. Oecologia (Berlin) 37: 285–295.

Simpson, R. L., R. E. Good, M. A. Leck, and D. F. Whigham. 1983. The ecology of freshwater tidal wetlands. BioScience 33: 255–259.

Sinicrope, T. L., R. Langis, R. M. Gersberg, M. J. Busnardo, and J. B. Zedler. 1992. Metal removal by wetland mesocosms subjected to different hydroperiods. Ecological Engineering 1: 309–322.

Sjörs, H. 1948. Myrvegetation i bergslagen. Acta Phytogeographica Suecica 21: 1–299.

Sklar, F. H., and W. H. Conner. 1979. Effects of altered hydrology on primary production and aquatic animal populations in a Louisiana swamp forest. In J. W. Day, Jr., D. D. Culley, Jr., R. E. Turner, and A. T. Humphrey, Jr. eds. *Proceedings of the 3rd Coastal Marsh and Estuary Management Symposium*, Louisiana State University, Division of Continuing Education, Baton Rouge, pp. 101–208.

Smemo, K. A., and J. B. Yavitt. 2006. A multi-year perspective on methane cycling in a shallow peat fen in central New York State, USA. Wetlands 26: 20–29.

Smith C. S., M. S. Adams, and T. D. Gustafson. 1988. The importance of belowground mineral element states in cattails (*Typha latifolia* L.), Aquatic Botany 30: 343–352.

Smith, T. J. 1987. Seed predation in relation to tree dominance and distribution in mangrove forests. Ecology 68: 266–273.

Smith, T. J. I., and W. E. Odum. 1981. The effects of grazing by snow geese on coastal salt marshes. Ecology 62: 98 106.

Smith, T. J., H-T. Chan, C. C. McIvor, and M. B. Robblee. 1989. Comparisons of seed predation in tropical, tidal forests on three continents. Ecology 70: 146–151.

Smith, T. J., K. G. Boto, S. D. Frusher, and R. L. Giddins. 1991. Keystone species and mangrove forest dynamics: The influence of burrowing by crabs on soil nutrient status and forest productivity. Estuarine, Coastal, and Shelf Science 33: 419–432.

Smith, T. J., M. B. Robblee, H. R. Wanless, and T. W. Doyle. 1994. Mangroves, hurricanes, and lightning strikes. BioScience 44: 256–262.

Smith, C. J., R. D. Delaune, and W. H. Patrick, Jr. 1982. Carbon and nitrogen cycling in a Spartina alterniflora salt marsh. In: J. R. Freney and I. E. Galvally, eds. The Cycling of Carbon, Nitrogen, Sulfur and Phosphorus in Terrestrial and Aquatic Ecosystems. Springer-Verlag, New York, pp. 97–104.

Spieles, D. J., and W. J. Mitsch. 2000a. The effects of season and hydrologic and chemical loading on nitrate retention in constructed wetlands: A comparison of low and high nutrient riverine systems. Ecological Engineering 14: 77–91.

Spieles, D. J., and W. J. Mitsch. 2000b. Macroinvertebrate community structure in high-and low-nutrient constructed wetlands. Wetlands 20: 716–729.

Spieles, D. J., and W. J. Mitsch. 2003. A model of macroinvertebrate trophic structure and oxygen demand in freshwater wetlands. Ecological Modelling 161: 183–194.

Stephenson, T. D. 1990. Fish reproductive utilization of coastal marshes of Lake Ontario near Toronto. Journal of Great Lakes Research 16: 71–81.

Steward, K. K. 1990. Aquatic weed problems and management in the eastern United States. In A. H. Pietersen, and K. J. Murphy, eds. *Aquatic Weeds: The Ecology and Management of Nuisance Aquatic Vegetation*. Oxford University Press, New York, pp. 391–405.

Stewart, G. R., and M. Popp. 1987. The ecophysiology of mangroves. In R. M. M. Crawford, ed. Plant Life in Aquatic and Amphibious Habitats, Spec Publ. British Ecological Society, vol. 5, pp. 333–345.

Stout, J. P. 1978. An Analysis of Annual Growth and Productivity of Juncus roemerianus Scheele and Spartina alterniflora Loisel in Coastal Alabama. Ph.D. Dissertation, University of Alabama, Tuscaloosa.

Strack, M., ed. 2008. *Peatlands and Climate Change*. International Peat Society, Jyvaskyla, Finland, 223 pp.

Stromberg, J. C. 2001. Influence of stream flow regime and temperature on growth rate of the riparian tree, *Platanus wrightii*, in Arizona. Freshwater Biology 46: 227–239.

Stuckey, R. L. 1980. Distributional history of *Lythrum salicaria* (purple loosestrife) in North America. Bartonia 47: 3–20.

Sullivan, M. J., and C. A. Currin. 2000. Community structure and functional dynamics of benthic microalgae in salt marshes. In M. PP. Weinstein and D. A. Kreeger, eds. *Concepts and Controversies in Tidal Marsh Ecology*. Kluwer Academic Publishers, The Netherlands.

Sullivan, T. J. 1993. Whole-ecosystem nitrogen effects research in Europe. Environmental Science & Technology 27: 1482–1486.

Sullivan, M. J., and C. A. Moncreiff. 1988. Primary production of edaphic algal communities in a Mississippi salt marsh. Journal of Phycology 24: 49–58.

Svengsouk, L. M., and W. J. Mitsch. 2001. Dynamics of mixtures of *Typha latifolia* and *Schoenoplectus tabernaemontani* in nutrient-enrichment wetland experiments. American Midland Naturalist 145: 309–324.

Swab, R. M., L. Zhang, and W. J. Mitsch. 2008. Effect of hydrologic restoration and *Lonicera maacki* removal on herbaceous understory vegetation in a bottomland hardwood forest. Restoration Ecology 16: 453–463.

Swarzenski, C., E. M. Swenson, C. E. Sasser, and J. G. Gosselink. 1991. Marsh mat flotation in the Louisiana Delta Plain, Journal of Ecology 79: 999–1011.

Sykes, P. W. Jr., C. B. Kepler, K. L. Litzenberger, H. R. Sansing, E. T. R. Lewis, and J. S. Hatfield. 1999. Density and habitat of breeding swallow-tailed kites in the lower Suwannee ecosystem, Florida. Journal of Field Ornithology 70: 321–336.

Szumigalski, A. R., and S. E. Bayley. 1996a. Net above-ground primary production along a bog-rich fen gradient in central Alberta, Canada. Wetlands 16: 467–476.

Szumigalski, A. R., and S. E. Bayley. 1996b. Decomposition along a bog to rich fen gradient in central Alberta, Canada. Canadian Journal of Botany 74: 573–581.

Tansley, A. G. 1935. The use and abuse of vegetational concepts and terms. Ecology 16: 284–307.

Taylor, J. R., M. A. Cardamone, and W. J. Mitsch. 1990. Bottomland hardwood forests: their functions and values. In J. G. Gosselink, L. C. Lee, and T. A. Muir, eds. *Ecological Processes and Cumulative Impacts: Illustrated by Bottomland Hardwood Wetland Ecosystems*. Lewis Publishers, Chelsea, MI, pp. 13–86.

Taylor, K. L., J. B. Grace, G. R. Guntenspergen, and A. L. Foote. 1994. The interactive effects of herbivory and fire on an oligohaline marsh, Little Lake, Louisiana, USA. Wetlands 14: 82–87.

Teal, J. M. 1962. Energy flow in the salt marsh ecosystem of Georgia, Ecology 43: 614–624.

Teal, J. M. 1986. The Ecology of Regularly Flooded Salt Marshes of New England: A Community Profile, U.S. Fish Wildl. Serv., Washington, DC, Biol. Repp. 85 (7.4), 61 pp.

Teal, J. M., and B. L. Howes. 2000. Salt marsh values: Retrospection from the end of the century. In M. P. Weinstein and D. A. Kreeger, eds. *Concepts and Controversies in Tidal Marsh Ecology*. Kluwer Academic Publishers, The Netherlands, pp. 9–19.

The Nature Conservancy. 1992. The Forested Wetlands of the Mississippi River: An Ecosystem in Crisis, The Nature Conservancy, Baton Rouge, LA, 25 pp.

Thibodeau, F. R., and N. H. Nickerson. 1986. Differential oxidation of mangrove substrate by *Avicennia germinas* and *Rhizophora mangle*. American Journal of Botany 73: 512–516.

Thom, B. G. 1982. Mangrove ecology: A geomorphological perspective. In Clough, B.F., ed. Mangrove Ecosystems in Australia. Australian National University Press, Canberra, pp. 3–17.

Thormann, M. N., and S. E. Bayley. 1997. Aboveground net primary productivity along a bog-fen-marsh gradient in southern boreal Alberta, Canada. Ecoscience 4: 374–384.

Thorp, J. H., E. M. McEwan, M. F. Flynn, and F. R. Hauer. 1985. Invertebrate colonization of submerged wood in a cypress-tupelo swamp and blackwater stream. American Midland Naturalist 113: 56–68.

Trettin, C. C., R. Laiho, K. Minkkinen, and J. Laine. 2006. Influence of climate change factors on carbon dynamics in northern forested peatlands. Canadian Journal of Soil Science 86: 269–280.

Turner, R. E. 1977. Intertidal vegetation and commercial yields of penaeid shrimp. American Fisheries Society Transactions 106: 411–416.

Turner, R. E., W. Woo, and H. R. Jitts. 1979. Estuarine influences on a continental shelf plankton community. Science 206: 218–220.

Tuttle, C. L., L. Zhang, and W. J. Mitsch. 2008. Aquatic metabolism as an indicator of the ecological effects of hydrologic pulsing in flow-through wetlands. Ecological Indicators 8: 795–806.

Twilley, R. R. 1982. Litter Dynamics and Organic Carbon Exchange in Black Mangrove (*Avicennia germinans*) Basin Forests in a Southwest Florida Estuary, Ph.D. dissertation, University of Florida, Gainesville, FL.

Twilley, R. R. 1985. The exchange of organic carbon in basin mangrove forests in a southwest Florida estuary, Estuarine, Coastal and Shelf Science 20: 543–557.

Twilley, R. R. 1998. Mangrove wetlands. In M.G. Messina and W.H. Conner, eds. Southern Forest Wetlands: Ecology and Management. Lewis Publishers, Boca Raton, FL, pp. 445–473.

Twilley, R. R., A. E. Lugo, and C. Patterson-Zucca. 1986. Litter production and turnover in basin mangrove forests in southwest Florida. Ecology 67: 670–683.

Twilley, R. R., R. H. Chen, and T. Hargis. 1992. Carbon sinks in mangroves and their implications to carbon budget of tropical coastal ecosystems. Water, Air, and Soil Pollution 64: 265–288.

Twilley, R. R., S. C. Snedaker, A. Yanez-Arancibia, and E. Medina. 1996. Biodiversity and ecosystem processes in tropical estuaries: Perspectives on mangrove ecosystems. In H. A. Mooney, J. H. Cushman, E. Medina, O. E. Sala, and E. D. Schulze, eds. *Functional Role of Biodiversity: A Global Perspective*. John Wiley & Sons, London, pp. 327–370.

Tyler, P. A. 1976. Lagoon of Islands, Tasmani—Deathknell for a unique ecosystem? Biological Conservation 9: 1–11.

Updegraff, K., S. D. Bridgham, J. Pastor, P. Weishampel, and C. Harth. 2001. Response of CO_2 and CH_4 emissions from peatlands to warming and water table manipulation. Ecological Applications 11: 311–326.

Urban, N. R., S. J. Eisenreich, and E. Gorham. 1985. Proton cycling in bogs: geographic variation in northeastern North America. In T. C. Hutchinson and

K. Meema, eds. *Proceedings NATO Advanced Research Workshop on the Effects of Acid Deposition on Forest, Wetland, and Agricultural Ecosystems*, Springer-Verlag, New York, pp. 577–598.

Urban, N. R., and S. J. Eisenreich. 1988. Nitrogen cycling in a forested Minnesota bog, Can J. Bot. 66: 435–449.

Valéry, L., V. Bouchard, and J. C. Lefeuvre. 2004. Impact of the invasive native species *Elymus athericus* on carbon pools in a salt marsh. Wetlands 24: 268–276.

van der Valk, A. G. 1981. Succession in wetlands: a Gleasonian approach. Ecology 62: 688–696.

van der Valk, A. G., and C. B. Davis. 1978. The role of seed banks in the vegetation dynamics of prairie glacial marshes. Ecology 59: 322–335.

van der Valk, A.G. 1994. Effects of prolonged flooding on the distribution and biomass of emergent species along a freshwater wetland coenocline. Plant Ecology 110: 185–196.

van der Valk, A. G., L. Squires, and C. H. Welling. 1994. Assessing the impacts of an increase in water level on wetland vegetation. Ecological Applications 4: 525–534.

Van Engel, W. A., and E. B. Joseph. 1968. Characterization of Coastal and Estuarine Fish Nursery Grounds as Natural Communities. Final Report, Bureau of Commercial Fisheries, Virginia Institute of Marine Science, Gloucester Point, VA, 43 pp.

Van Raalte, C. D. 1976. Production of epibenthic salt marsh algae: light and nutrient limitation. Limnology and Oceanography 21: 862–872.

Verhoeven, J. T. A. 1986. Nutrient dynamics in mincrotrophic peat mires, Aquatic Botany 25: 117–137.

Verhoeven, J. T. A., D. F. Whigham, M. van Kerkhoven, J. O'Neill, and E. Maltby. 1994. Comparative study of nutrient-related processes in geographically separated wetlands: Toward a science base for functional assessment procedures. In Mitsch, W. J., ed. *Global Wetlands: Old World and New*. Elsevier, Amsterdam, pp. 91–106.

Verhoeven, J. T. A., D. F. Whigham, R. van Logtestijn, and J. O'Neill. 2001. A comparative study of nitrogen and phosphorus cycling in tidal and non-tidal river wetlands. Wetlands 21: 210–222.

Vermeer, J. G., and F. Berendse. 1983. The relationship between nutrient availability, shoot biomass, and species richness in grassland and wetland communities. Vegetatio 53: 121–126.

Vernon, R. O. 1947. Cypress domes. Science 105: 97–99.

Visser, J. M. 1989. The Impact of Vertebrate Herbivores on the Primary Production of *Sagittaria* Marshes in the Wax Lake Delta, Atchafalaya Bay, Louisiana. Ph.D. dissertation, Louisiana State University, Baton Rouge, 88 pp.

Visser, J. M., C. E. Sasser, R. G. Chabreck, and R. G. Linscombe. 1998. Marsh vegetation types of the Mississippi River deltaic plain. Estuaries 21: 818–828.

Visser, J. M., C. E. Sasser, R. H. Chabreck, and R. G. Linscombe. 1999. Long-term vegetation change in Louisiana tidal marshes. Wetlands 19: 168–175.

von Post, L., and E. Granlund. 1926. Södra Sveriges Torvtillgängar. Sveriges Geologiska Undersökning Ser. C Avhandlingar och uppsater, No. 355. Arsbok 19 (2): 1–127.

Waddington, J. M., M. J. Greenwood, R. M. Petrone, and J. S. Price. 2003. Mulch decomposition impedes recovery of net carbon sink function in a restored peatland. Ecological Engineering 20: 199–210.

Waddington, J. M., K. Tóth, and R. Bourbonniere. 2008. Dissolved organic carbon export from a cutover and restored peatland. Hydrologic Processes 22: 2215–2224.

Wadsworth, J. R., Jr. 1979. Duplin River Tidal System. Sapelo Island, Georgia. Map reprinted in Jan. 1982 by the University of Georgia Marine Institute, Sapelo Island, GA.

Wafer, S., A.G. Untawale, and M. Wafar. 1997. Litter fall and energy flux in a mangrove ecosystem. Estuarine, Coastal and Shelf Science 44: 111–124.

Wainscott, V. J., C. Bardey, and P. P. Kangas. 1990. Effect of muskrat mounds on microbial density on plant litter. American Midland Naturalist 123: 399–401.

Walter, H. 1973. *Vegetation of the Earth*, Springer-Verlag, New York, 237 pp.

Wang, N., and W. J. Mitsch. 2000. A detailed ecosystem model of phosphorus dynamics in created riparian wetlands. Ecological Modelling 126: 101–130.

Ward, G. A., T. J. Smith III, K. R. T. Whalen, and T. W. Doyle. 2006. Regional process in mangrove ecosystems: spatial scaling relationships, biomass, and turnmover rates following catastrophic disturbance. Hydrobiologia 569: 517–527.

Warner, B. G., and C. D. A. Rubec, eds. 1997. *The Canadian Wetland Classification System*. National Wetlands Working Group, Wetlands Research Centre, University of Waterloo, Waterloo, Ontario, Canada.

Warner, D., and D. Wells. 1980. *Bird Population Structure and Seasonal Habitat Use as Indicators of Environment Quality of Peatlands*, MN Department of Natural Resources, St. Paul, MN, 84 pp.

Watzin, M. C., and J. G. Gosselink. 1992. Coastal Wetlands of the United States. Louisiana Sea Grant College Program, Baton Rouge, LA and U.S. Fish and Wildlife Service, Lafayette, LA, 15 pp.

Weber, C. A. 1907. Aufbau und Vegetation der Moore Norddutschlands. Beibl. Bot. Jahrb. 90: 19–34.

Weihe, P. E., and W. J. Mitsch. 2000. Garden wetland experiment demonstrates genetic differences in soft rush obtained from different regions (Ohio). Ecological Restoration 18: 258–259.

Weller, M. W., and C. S. Spatcher. 1965. Role of habitat in the distribution and abundance of marsh birds, Iowa State University Agric. and Home Economics Exp. Station Spec. Report 43, Ames, IA, 31 pp.

Weller, M. W. 1994. Freshwater Marshes, 3rd ed. University of Minnesota Press, Minneapolis, MN, 192 pp.

Weller, M. W. 1999. Wetland Birds. Cambridge University Press, Cambridge, UK.

Weltzin, J. F., J. Pastor, C. Harth, S.D. Bridgham, K. Updegraff, and C. T. Chapin. 2000. Response of bog and fen plant communities to warming and water-table manipulations. Ecology 81: 3464–3478.

Weltzin, J. F., J. K. Keller, S. D. Bridgham, J. Pastor, P. B. Allen, and J. Chen. 2005. Litter controls plant community composition in a northern fen. Oikos 110: 537–546.

Werme, C. E. 1981. Resource Partitioning in the Salt Marsh Fish Community, Ph.D. dissertation, Boston University, Boston, 126 pp.

Werner, K. J., and J. B. Zedler. 2002. How sedge meadow soils, microtopography, and vegetation respond to sedimentation. Wetlands 22: 451–466.

Wharton, C. H., H. T. Odum, K. Ewel, M. Duever, A. Lugo, R. Boyt, J. Bartholomew, E. DeBellevue, S. Brown, M. Brown, and L. Duever. 1976. *Forested Wetlands of Florida — Their Management and Use*, Center for Wetlands, University of Florida, Gainesville, 421 pp.

Wharton, C. H., W. M. Kitchens, E. C. Pendleton, and T. W. Sipe. 1982. The ecology of bottomland hardwood swamps of the Southeast: a community profile, U.S. Fish and Wildlife Service, Biological Services Program FWS/OBS-81/37, 133 pp.

Wheeler, B. D. 1980. Plant communities of rich fen systems in England and Wales. Journal of Ecology 68: 365–395.

Wheeler, B. D., and K. E. Giller. 1982. Species richness of herbaceous fen vegetation in Broadland, Norfolk, in relation to the quantity of above-ground plant material, Journal of Ecology 70: 179–200.

Whigham, D. F., and R. L. Simpson. 1975. Ecological Studies of the Hamilton Marshes, Progress report for the period June 1974–January 1975, Rider College, Biology Department, Lawrenceville, NJ.

Whigham, D. F., and C. J. Richardson. 1988. Soil and plant chemistry of an Atlantic white cedar wetland on the Inner Coastal Plain of Maryland. Canadian Journal of Botany 66: 568–576.

Whigham, D. F., R. L. Simpson, and K Lee. 1980. The Effect of Sewage Effluent on the Structure and Function of a Freshwater Tidal Wetland. Water Resources Research Institute Report, Rutgers University, New Brunswick, NJ, 160 pp.

White, D. A. 1993. Vascular plant community development on mudflats in the Mississippi River delta, Louisiana, USA. Aquatic Bot. 45: 171–194.

White, J. R., K. R. Reddy, and M. Z. Moustafa. 2004. Influence of hydrologic regime and vegetation on phosphorus retention in Everglades stormwater treatment area wetlands. Hydrological Processes 18: 343–355.

White, J. R., K. R. Reddy, and J. Majer-Newman. 2006. Hydrologic and vegetation effects on water column phosphorus in wetland mesocosms. Soil Science Society of America Journal 70: 1242–1251.

White, D. A., T. E Weiss, J. M. Trapani, and L. B. Thien. 1978. Productivity and decomposition of the dominant salt marsh plants in Louisiana. Ecology 59: 751–759.

Wieder, R. K., and G. E. Lang. 1983. Net primary production of the dominant bryophytes in a Sphagnum-dominated wetland in West Virginia, Biologist 86: 280–286.

Wieder, R. K., D. H. Vitt, and B. W. Benscoter. 2006. Peatlands and the boreal forest. In R. K. Wieder and D. H. Vitt, eds., *Boreal Peatland Ecosystems*, Springer, Heiderberg. pp. 1–8.

Wiegert, R. G. 1986. Modeling spatial and temporal variability in a salt marsh: sensitivity to rates of primary production, tidal migration and microbial degradation. In D. A. Woffe, ed. *Estuarine Variability*. Academic Press, New York, pp. 405–426.

Wiegert, R. G., R. R. Christian, J. L. Gallagher, J. R. Hall, R. D. H. Jones, and R. L. Wetzel. 1975. A preliminary ecosystem model of a Georgia salt marsh. In L. E. Cronin, ed. *Estuarine Research Vol. 1*. Academic Press, New York, pp. 583–601.

Wiegert, R. G., and R. L. Wetzel. 1979. Simulation experiments with a fourteen compartment model of a Spartina salt marsh, in R. F. Dame, ed. *Marsh-Estuarine Systems Simulations*. University of South Carolina Press, Columbia, pp. 7–39.

Wiegert, R. G., R. R. Christian, and R. L. Wetzel. 1981. A model view of the marsh. In L. R. Pomeroy and R. G. Wiegert, eds. *The Ecology of a Salt Marsh*. Springer-Verlag, New York, pp. 183–218.

Wiegert, R. G., and B. J. Freeman. 1990. Tidal Salt Marshes of the Southeast Atlantic Coast: A Community Profile, U. S. Department of Interior, Fish and Wildl. Service, Biological Report 85 (7.29), Washington, DC.

Willis, C., and W. J. Mitsch. 1995. Effects of hydrologic and nutrients on seedling emergence and biomass of aquatic macrophytes from natural and artificial seed banks. Ecological Engineering 4: 65–76.

Wilson, J. O., R. Buchsbaum, I. Valiela, and T. Swain. 1986. Decomposition in salt marsh ecosystems: Phenolic dynamics during decay of litter of *Spartina alterniflora*. Marine Ecology Progress Series 29: 177–187.

Wozniak, J. R., D. L. Childers, W. T. Anderson, D. T. Rudnick, and C. J. Madden. 2008. An in situ mesocosm methods for quantifying nitrogen cycling rates in oligotrophic wetlands using 15N tracer techniques. Wetlands 28: 502–512.

Wray, H. E., and S. E. Bayley. 2007. Denitrification rates in marsh fringes and fens in two boreal peatlands in Alberta, Canada. Wetlands 27: 1036–1045.

Wright, R. B., B. G. Lockaby, and M. R. Walbridge. 2001. Phosphorus availability in an artificially flooded southeastern floodplain forest soil. Soil Science Society of America Journal 65: 1293–1302.

Wu, X., and W. J. Mitsch. 1998. Spatial and temporal patterns of algae in newly constructed freshwater wetlands. Wetlands 18: 9–20.

Wu, Y., F. H. Sklar, K. Gopu, and K. Rutchey. 1996. Fire simulations in the Everglades Landscape using parallel programming. Ecological Modelling 93: 113–124.

Yánez-Arancibia, A., A. L. Lara-Dominguez, J. L. Rojan-Galaviz, P. Sánchez-Gil, J. W. Day, and C. J. Madden. 1988. Seasonal biomass and diversity of estuarine fishes coupled with tropical habitat heterogeneity (southern Gulf of Mexico). Journal of Fish. Biology 33 (Suppl. A): 191–200.

Yánez-Arancibia, A., A. L. Lara-Dominguez, and J. W. Day. 1993. Interactions between mangrove and seagrass habitats mediated by estuarine nekton assemblages: Coupling of primary and secondary production. Hydrobiologia 264: 1–12.

Yarbro, L. A., 1983. The influence of hydrologic variations on phosphorus cycling and retention in a swamp stream ecosystem. In T. D. Fontaine and S. M. Bartell, eds. *Dynamics of Lotic Ecosystems*. Ann Arbor Science, Ann Arbor, MI, pp. 223–245.

Zavalishin, N. N. 2008. Dynamic compartment approach for modelling regimes of carbon cycle functioning in bog ecosystems. Ecological Modelling 213: 16–32.

Zedler, J. B. 1980. Algae mat productivity: Comparisons in a salt marsh. Estuaries 3: 122–131.

Zhang, L., and W. J. Mitsch. 2005. Modelling hydrological processes in created wetlands: An integrated system approach. Environmental Modelling & Software 20: 935–946.

Zhang, L., and W. J. Mitsch. 2007. Sediment chemistry and nutrient influx in a hydrologically restored bottomland hardwood forest in Midwestern USA. River Research and Applications 23: 1026–1037.

Zhang, Y., C. S. Li, C. C. Trettin, H. Li, and G. Sun. 2002. An integrated model of soil, hydrology, and vegetation for carbon dynamics in wetland ecosystems. Global Biogeochemical Cycles 16: 1029–1061.

Zhu, R., Y. Liu, J. Ma, H. Xu, and L. Sun. 2008. Nitrous oxide flux to the atmosphere from two coastal tundra wetlands in eastern Antarctica. Atmospheric Environment 42: 2437–2447.

Ziser, S. W. 1978. Seasonal variations in water chemistry and diversity of the phytophilia macroinvertebrates of three swamp communities in southeastern Louisiana. Southwest Naturalist 23: 545–562.

Zoltai, S. C. 1988. Wetland environments and classification. In *National Wetlands Working Group*, Wetlands of Canada, Ecological Land Classification Series, no. 24. Environment Canada, Ottawa, Ontario, and Polyscience Publications, Inc., Montreal, Quebec, pp. 1–26.

General Index

biogeochemistry, 163–166,
184–187
carbon, 13, 15, 185–187
classification, 157–163
consumers, 174–176
decomposition, 179–181
energy flow, 182–183
geographical extent, 151–153
hydrology and development,
153–157
mathematical models, 233–234
mesocosms, 194, 202–204
nutrient budgets, 184–187
peat accumulation, 181
primary productivity, 175, 177–180
restoration, 213–216
vegetation, 166–174
whole-ecosystem experiment,
213–216
Pennsylvania, freshwater swamps, 91
Philippines, mangroves, 63
Phosphorus
freshwater marshes, 122, 129,
143–147, 225–228
freshwater swamps, 98–99, 111,
116–118
mangrove wetlands, 86
mesocosm experiments, 192,
194–195, 199, 201
models, 217–218, 225–228
peatlands, 161, 166, 169, 172–173,
181
tidal salt marsh model, 217–218
whole-ecosystem wetland
experiments, 206, 209
Pneumatophores
cypress, 104–106
mangroves, 75–76
Pocosins, 151, 153, 170–171. *See also*
Peatlands
Poland
mesocosm, 200–201
peatlands, 154
Prairie potholes, 9, 118
Primary productivity
freshwater marshes, 136, 138–139,
213–214
freshwater swamps, 110–116
mangrove wetlands, 78–80
peatlands, 175, 177–180
tidal freshwater wetlands,
56–58
tidal salt marshes, 38–41

Prop roots, mangroves, 75–76
Puerto Rico, mangroves, 69, 79–80

Quaking bog, 154–155
Quebec
peatlands, 177, 180, 213–216
whole-ecosystem wetland
experiment, 205, 213–216

Raised bog, 154–158, 161, 164, 177.
See also Ombrotrophic bog;
Peatlands
concentric, 157, 159
excentric, 157, 159
Ramsar Convention, 3
Red maple swamp, 91, 96, 102
Redfield ratio, 145
Reed swamp, 88, 90
die-back, 127
freshwater marshes, 131
Reptiles
freshwater swamps, 109
peatlands, 174–175
tidal freshwater marshes, 55–56
Restoration
peatlands, 213–216
bottomland hardwood forest,
211–213
Rheotrophic peatlands, 160
Riparian forested wetlands, 9, 93,
96–97
area, 3
energy flow diagram, 12
River flood mitigation, 16
River floodplain
mesocosm, 195, 200–201
restoration, 211–213
Riverine mangroves, 67–68
Russia, 1
peatlands, 151, 177, 182–183, 233
salt marshes, 23, 26

Salinity control, mangroves, 75–76
Salinity
mangrove, 69–71
ocean, 20–21
salt marshes, 29–30
tidal salt marshes, 21
Salt exclusion, 76
Salt marshes, *see* Tidal salt marshes
Salt secretion, 76
Salton Sea, California, 123
Sand barrens, 29

Saskatchewan
freshwater marshes, 121
peatlands, 233
Scandinavia, salt marshes, 23
Schwingmoor, 154. *See also* Quaking
bog
Sea level rise
effect on mangroves, 82
modeling effect on coastal wetlands,
220
Sediment accretion, salt marshes, 41
Seed banks
freshwater marshes, 127
tidal freshwater wetlands, 50–51
Seed dispersal, forested wetlands, 106
Seed germination
cypress, 106
freshwater marshes, 127
Semi-anadromous fish, 52–53
Senegal, mangroves, 8, 75
Sloughs, 96
Soils
mangrove swamps, 69–71
salt marshes, 29–30
Soligenous peatland, 161–162
South Carolina
freshwater marshes, 138
freshwater swamps, 104
freshwater swamps, 106, 110, 112
116–117
pocosins, 170
salt marshes, 40
tidal freshwater swamps, 58
wetland mathematical model, 223
South Dakota, freshwater marshes,
121
Spatial models, Florida Everglades,
231–233
Species diversity
freshwater marshes, 129
peatlands, 185
STELLA simulation language,
224–225
Strings, peatlands, 157, 160
Succession
allogenic, 74
autogenic, 74
mangroves, 74
peatlands, 154–157
Sud, Africa, floating marshes, 49
Sulfate reduction, tidal swamps, 59–60

Organism Index

Acer (maple), 99
    rubrum (Red maple), 9, 49, 87, 91,
      93, 98, 100, 102, 107, 109, 171
    saccharinum (Silver maple), 102
Acipenser
    brevirostrum (Shortnose Sturgeon),
      53
    oxyrhynchus (Atlantic Sturgeon), 53
Acnida cannabina, 48
Acorus calamus (Sweet flag), 47, 57,
    138
Acris crepitans blanchardii
    (Blanchard's cricket frog), 137
Adinia xenica (Diamond Killifish), 37
Agelaius phoeniceus (Red-winged
    Blackbird), 36, 55, 134, 176, 225
Agkistrodon piscivorus, (Cottonmouth,
    Water moccasin), 55, 109
Aglaius phoniceus (Red-winged
    Blackbird), 134
Agrostis stolonifera, 25
Aix sponsa (Wood Duck), 54, 134,
    176
Ajaia ajaja (Roseate Spoonbill), 78
Alces alces (moose), 174
Alder (*Alnus*), 107
Alisma plantago-aquatica (Water
    plantain), 102
Alkali bulrush (*Scirpus robustus*), 126
Allenrolfea, 26
Alligator mississippiensis (American
    alligator), 109

Alligator weed (*Alternanthera
    philoxeroides*), 128
Alnus (alder), 107
    rugosa (Speckled alder), 103, 167
Alosa (Herring), 53
    mediocris (Hickory Shad), 53
Alternanthera philoxeroides (Alligator
    weed), 128
Amaranthus cannabinus (Water
    hemp), 46, 57
Amblystegiaceae, 169
Ambrosia (Ragweed), 47, 123
    trifida (Giant ragweed), 46, 57
Ambystoma
    jeffersonianum (Jefferson's
      salamander complex), 136–137
    maculatum (spotted salamander),
      136–137
    opacum (marbled salamander), 137
    talpoideum (mole salamander), 174
    texanum (smallmouth salamander),
      136–137
    tigrinum (tiger salamander), 137
American alligator (*Alligator
    mississippiensis*), 55, 109
American Bittern (*Botaurus
    lentiginosus*), 134
American bullfrog (*Rana catesbeiana*),
    137
American Coot (*Fulica americana*),
    134

American eel (*Anguilla rostrata*), 52,
    53
American elm (*Ulmus americana*), 91,
    102
American Goldfinch (*Carduelis
    tristis*), 55
American hornbeam (*Carpinus
    caroliniana*), 49
American Kestrel (*Falco sparverius*),
    55
American lotus (*Nelumbo lutea*), 226
American Redstart (*Sotophaga
    rutioilla*), 176
American toad (*Bufo americanus*),
    109, 135, 137
American Wigeon (*Anas americana*),
    36
American Woodcock (*Scolopax
    minor*), 55
Amia calva (bowfin), 52, 109
Ammodramus
    caudacutus (Sharptailed Sparrow),
      36
    maritimus (Seaside Sparrow), 36
Ammospiza caudacuta (Sharp-tailed
    sparrow), 175
Amphipod, 37, 42–43, 84, 108, 132
Anas
    americana (American Wigeon), 36,
      133
    clypeata (Northern Shoveler), 133
    crecca (Green-winged Teal), 36, 141